The Codex and Craft
in Late Antiquity

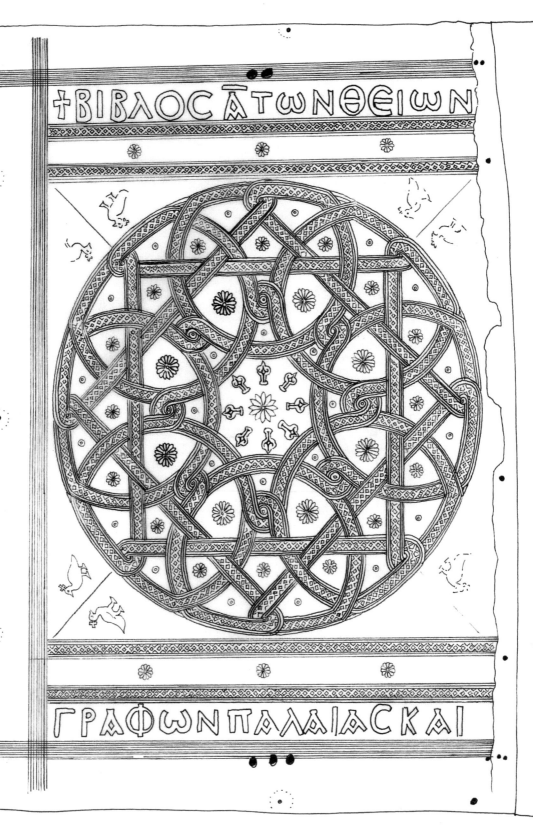

†ΒΙΒΛΟСᾹΤΩΝΘΕΙΩΝ

ΓΡΑΦΩΝΠΑΛΛΑΙΑСΚΑΙ

The Codex and Crafts in Late Antiquity

Georgios Boudalis

Bard Graduate Center
New York City

This catalogue is published in conjunction with the exhibition *The Codex and Crafts in Late Antiquity* held at Bard Graduate Center Gallery from February 23 through July 8, 2018.

Curator of the Exhibition: Georgios Boudalis

Focus Project Team:
Director of the Gallery and Executive Editor, Gallery Publications: Nina Stritzler-Levine
Head of Focus Gallery Project: Ivan Gaskell
Chief Curator: Marianne Lamonaca
Associate Curator: Caroline Hannah
Director of Publishing: Daniel Lee
Catalogue Design and Production: Designed by Jocelyn Lau with Art Direction by Kate DeWitt
Director of the Digital Media Lab: Jesse Merandy
Manager of Rights and Reproductions: Alexis Mucha

Published by Bard Graduate Center, New York City

Typeface: This book is set in GT Sectra Display and Graphik.

Exclusive trade distribution by The University of Chicago Press, Chicago and London

ISBN: 978-1-941-79212-4

Cover: Ethiopic 8, 17th–18th century, Carolina Rediviva Library, Uppsala, Sweden. Frontispiece: Interlaced pattern tooled on right board cover, Syriac codex Ambrosianus Syriacus C 313 inf. Back cover: Inside of upper cover originally on MS M.577, Egypt, 9th–10th century AD. The Morgan Library and Museum, Purchased for J. Pierpont Morgan, 1911, MS M.577A1.

Library of Congress Cataloging-in-Publication Data

Names: Boudalis, Georgios.
Title: The codex and crafts in late antiquity / Georgios Boudalis.
Description: New York City : Bard Graduate Center ; Chicago : Exclusive trade distribution by The University of Chicago Press, [2017] | This catalogue is published in conjunction with the exhibition The Codex and Crafts in Late Antiquity held at Bard Graduate Center Gallery February 23–June 24, 2018.
Identifiers: LCCN 2017039257 | ISBN 9781941792124 (paperback)
Subjects: LCSH: Books—Byzantine Empire—History—To 400—Exhibitions. | Books—Byzantine Empire—History—400–1400—Exhibitions. | Manuscripts, Byzantine—Exhibitions. | Bookbinding, Medieval—Byzantine Empire—Exhibitions. | Byzantine antiquities—Exhibitions. | BISAC: SOCIAL SCIENCE / Archaeology. | HISTORY / Ancient / General.
Classification: LCC Z8.B9 B68 2017 | DDC 090.74—dc23
LC record available at https://lccn.loc.gov/2017039257

10 9 8 7 6 5 4 3

Printed in the United States of America.

Unless otherwise noted, all drawings in this book are by the author.

To Mum and Dad

Contents

Director's Foreword

Books are things of the past. Some make this claim because they believe that electronic media have superseded books. But the sentence is ambiguous. In its other sense, books are things that can inform us about earlier times through their particular materiality. It is in this manner that the books studied by Georgios Boudalis are significant. Although centuries old, the codices he investigates are also things of the present—survivals of a bygone era, they are nevertheless works of inventive craft from which we can learn about how innovation takes place.

Boudalis is a scholar of ancient books whose work is firmly grounded in technical examination. He studied fine arts in Thessaloniki and art conservation in Florence and Athens, specializing in books. He joined the conservation staff of the Museum of Byzantine Culture in Thessaloniki in 2000 and is now the head of its Book and Paper Conservation Laboratory. He completed his PhD at the University of the Arts in London in 2005, with a dissertation on the evolution of the Byzantine bookbinding tradition from the end of the Byzantine Empire until the early eighteenth century. Since then, he has specialized in analyzing the fabric of early Byzantine codices.

While he was a research fellow at Bard Graduate Center in spring 2015, Boudalis made such an impression that we invited him to return as a visiting adjunct professor for the fall of 2016. During that semester, he continued to work on how those who developed the new craft of making codices in the Eastern Mediterranean of late antiquity assimilated existing craft skills and adapted them to new ends. Byzantine codices are no less significant as examples of the decorative arts of late antiquity than are socks, sandals, and baskets, the making of which required comparable finesse.

Books are things too. Their gatherings, endbands, and bindings required sewing, looping, and decorative tooling. Boudalis shows us precisely how these skills found new expression as the codex superseded the roll (or scroll) as the principal textual vehicle in late antiquity—a development that echoes to this day in the form of the books with which we are familiar, books that are anything but things of the past.

To prepare *The Codex and Crafts of Late Antiquity*, Boudalis taught an In Focus course in the fall semester 2016, "The Making of the Early Codex and the Crafts of Late Antiquity." The students who participated in the class and contributed to the project were Jaime Ding, Emily Field, Julia Lillie, Aleena Malik, Sarah Reetz, and Darienne Turner, the last of whom, as a curatorial fellow, also served ably as Boudalis's research assistant from spring 2016 through spring 2017.

The author conducted much of his research for this project in New York at the Morgan Library and Museum, drawing on its late antique collections. Generous loans from this institution have made the exhibition possible, and I should like to thank its director, Colin B. Bailey. We are also grateful to all the other lenders to the exhibition: the Brooklyn Museum and its director, Anne Pasternak; the Metropolitan Museum of Art and its director at the time of the approval of the loan requests, Thomas P. Campbell; the University of Pennsylvania Museum of Archaeology and Anthropology and its director, Julian Siggers; the American Museum of Natural History and the chair of its Division of Anthropology, Laurel Kendall; and two museums at Yale University: the Yale Peabody Museum of Natural History and its director, David K. Skelly; the Yale University Art Gallery and its director, Jock Reynolds; and the makers of replicas and facsimiles, Regina de Giovanni, Ursula Mitra, and Georgios Boudalis.

The Focus Project is a collaboration between the Gallery and the Degree Programs and Research Institute of the Bard Graduate Center. Peter Miller, our dean, and Nina Stritzler-Levine, director of the Gallery and executive editor of BGC Gallery Publications, joined forces to make it possible. Elena Pinto Simon, dean of Academic Administration and Student Affairs, provided vital advice at all stages until her departure in June 2016. The head of the Focus

Project, who supervised all aspects of *The Codex and Crafts of Late Antiquity*, is Ivan Gaskell, professor in our Degree Programs and Research Institute.

Many professional staff of the Degree Programs and Research Institute and the Gallery contributed to the realization of Boudalis's concept. They include Eric Edler, Gallery registrar; Kate Dewitt, art director; Caroline Hannah, associate curator; Marianne Lamonaca, associate Gallery director and chief curator; Jocelyn Lau, junior designer; Daniel Lee, director of publishing; Jesse Merandy, director of the Digital Media Lab; Alexis Mucha, manager of rights and reproductions; and Emily Reilly, director of public programs. Jeremy Johnston of Darling Green created a sensitive display for the exhibtion.

Our copyeditor, Carolyn Brown, and our proofreader, Christine Gever, worked with great care to prepare this publication for the press. My thanks go to all concerned among the faculty and staff of Bard Graduate Center whose care has brought *The Codex and Crafts of Late Antiquity* to completion.

— Susan Weber
Director and Founder
Iris Horowitz Professor in the History of the Decorative Arts
Bard Graduate Center

Foreword

Writing is only used in some societies and at certain times. Western people have long paid a great deal of attention to the written word, many assuming—quite wrongly—that it alone permits sophisticated mental activity, a contention decisively refuted by various anthropologists in the first half of the twentieth century.[1] Although, over time, more Westerners have been illiterate than literate, the written word is a key instrument of power in Western and various other societies. Cultural guardians use it to transmit lore and knowledge across space and time, while governments use it to distribute orders and decisions and to maintain taxation records, thereby allowing complex state formation. As political scientist and anthropologist James C. Scott has shown, it is no wonder that those who wish to escape or evade large-scale hierarchical social structures eschew or renounce not only visible crops subject to state taxation but also writing as the technology that makes subjugation through taxation possible.[2]

Even though linguist Milman Parry demonstrated in 1928 that, before they were written down, the Homeric epics had been transmitted orally, scholars of the Greater Mediterranean have long given almost exclusive attention to the written word.[3] In his turn, Parry's student, Albert Lord, brought attention to the continuation of oral tradition into the present in the Greater Mediterranean world, following Parry in recording epic recitations in the Balkans and elsewhere.[4] Thanks to Parry's work, classical scholar Eric Havelock proposed a radical break in modes of thought and expression between the pre-Socratic Greek philosophers, whom he identified as working in an oral poetic mode, and Plato, who in the *Republic* famously rejected poetry, setting the scene for writing rather than speech as the most authoritative mode of linguistic expression.[5] In spite of the survival of some Mediterranean oral literature—such as the stories familiar as Aesop's fables—before and after it was recorded in writing, the written word has ruled in the Mediterranean, particularly in the Greek, Roman, and Byzantine worlds, for nearly three thousand years, and in Egypt for about two thousand years longer.

Writing as a technology, as distinct from pictograms, depends on relating human linguistic sounds to visible marks in sequence. The alphabetical system of the Greek, Roman, and Byzantine worlds, in which each sign represents a phoneme (the system with which Westerners are familiar), is just one among several distinct modes of writing. In worldwide terms, alphabetical writing systems have clearly been in the minority. Many written languages conform to other systems, among them some of those most widely used in human history, such as Arabic (in which each sign represents a consonant) and Chinese (in which each character represents a syllable that may have more than one pronunciation and meaning).[6]

Regardless of the form of writing, as soon as writers learned to make marks with an agent such as a liquid that leaves a trace when it dries (ink or paint), they required not only a suitable surface on which to make those marks but also a means of preserving and ordering their inscriptions that exceeded the extent of a single surface that could be made visible all at once. The earliest solution in the Mediterranean world was the scroll, which allowed the user to roll and unroll written material between two cylinders, progressively revealing part of its surface. The scroll was a technical response to the physical character of the most common writing support in this part of the world: thin, pliable sheets made from the pith of the papyrus plant. Papyrus, although weaker than parchment, was commonly folded to make the earliest single-gathering and multigathering codices. Parchment or leather stays were used inside or outside the fold to reinforce the papyrus in single-gathering codices. As Georgios Boudalis shows, when prepared animal skins (parchment) began to replace papyrus in the Byzantine world of late antiquity, codices became more flexible and resilient.

The codex gradually superseded the scroll, in part because it had mechanical advantages in terms of ease of consultation, especially in the case of long texts.[7]

Each codex that Boudalis studies contains a text that can be thought of as having an existence independent of any single physical instantiation. Even though they may take bibliographical considerations into account, most scholars are preoccupied with that text as matter for interpretation. In contrast, Boudalis invites us to consider not the text but the material thing that embodies it. He makes the key point that the birth of the codex depended on the adaptation of skills employed in other crafts. Makers facing the new challenge of producing codices readily adapted techniques already in use in making socks with cross-knit looping, sewing shoes, and weaving baskets. Whether this local claim regarding how innovation proceeded might constitute a general principle of technical innovation is an interesting philosophical question. Is each technological change dependent on the creative adaptation of an existing means from another sphere? Even if this is not invariably the case, there is a tendency in much Western thought to pay attention to seemingly ex nihilo invention while ignoring adaptive continuities. Boudalis's study prompts us to regard, by a process of lateral thinking, the adaptation of existing skills and methods for new purposes as being as worthy of attention as allegedly pure invention.

In introducing Georgios Boudalis's work, I have thought it important to invite attention to the contingent circumstances of the steps that craftspeople of late antiquity took by reminding readers that other technologies of both script itself and its embodiment exist elsewhere in the world, and that many societies—though proportionally fewer than in earlier times—have functioned without any such technologies. Even so momentous a development as the codex, the form of which remains in use to this day, should not necessarily be seen as of equal importance to all humans. In seeking to establish a context of human equity for the fascinating story of *The Codex and Crafts in Late Antiquity*, however, it would be willful—even foolish—not to acknowledge the contemporary dominant circumstances of published text production. As its reader, you hold a direct descendent of a Byzantine-era codex in your hands. But there are differences between such codices and this book. In this digital age, the traces with which this text was formed were originally purely electronic, a mode of production inconceivable to the writers and binders of the codices presented in these pages.

— Ivan Gaskell
Professor of Cultural History and Museum Studies
Curator and Head of the Focus Gallery Project
Bard Graduate Center

Notes

1. Prominent among anthropologists' accounts of sophisticated cosmologies transmitted orally is Marcel Griaule's influential *Dieu d'eau: Entretiens avec Ogotemmêli* (Paris: Éditions du Chêne, 1948), although his contentions have subsequently not escaped criticism. See, for example, Walter E. A. van Beek, "Dogon Revisited: A Field Evaluation of the Work of Marcel Griaule," *Current Anthropology* 32 (1991): 139–158, and comments and responses that immediately follow his article by Suzanne Preston Blier, Jacky Bouju, Peter Ian Crawford, Mary Douglas, Paul Lane, and Claude Meillassoux (158–167).

2. See James C. Scott, *The Art of Not Being Governed: An Anarchist History of Upland Southeast Asia* (New Haven, Conn.: Yale University Press, 2009).

3. See *The Making of Homeric Verse: The Collected Papers of Milman Parry*, edited by Albert Parry (Oxford: Clarendon, 1971).

4. Albert B. Lord, *The Singer of Tales* (Cambridge, Mass.: Harvard University Press, 1960), and *Epic Singers and Oral Tradition* (Ithaca, N.Y.: Cornell University Press, 1991). The Milman Parry Collection of Oral Literature is in the Widener Library of Harvard University, http://chs119.chs.harvard.edu/mpc/.

5. Eric A. Havelock, *Preface to Plato* (Cambridge, Mass.: Belknap Press of Harvard University Press, 1963).

6. See Florian Coulmas, *Writing Systems: An Introduction to Their Linguistic Analysis* (New York: Cambridge University Press, 2003).

7. Much has been written on the origins of the codex; see, among others, Colin H. Roberts and T. C. Skeat, *The Birth of the Codex* (New York: Oxford University Press for the British Academy, 1983), which stresses the importance of the codex in the development of Christianity.

Acknowledgments

In September 2012 the Center for the Study of the Material Text in Cambridge, UK, organized "Texts and Textiles," a conference at which I presented a paper titled "From Fabric to Bookbinding: The Technological Background of the Codex Structure." I still remember vividly the period when I was pursuing the research for that paper, not only because of the excitement that accompanied every single small discovery but also because at the time I was struggling to grasp a field completely unknown to me, that of fabrics and fabric making. The results of that research revealed to me a much wider field of all the possible connections one could identify between the making of the codex and the different crafts of late antiquity, the period when the codex—that is, the book as we know it today—first took shape and was gradually established as the standard form of the book, supplanting the roll. The opportunity to further pursue that research arose in the best possible way when Bard Graduate Center announced a number of scholarships for research related to material culture, to be conducted at the Center in Manhattan. I was fortunate to be granted a scholarship to continue my research and subsequently spent four very productive months in 2015 between Bard Graduate Center, the Morgan Library and Museum, and several other New York City museums and libraries. But that was only the beginning, for Dean Peter Miller then asked me whether I would like to return to BGC to teach a class in the fall semester of 2016 and to curate a February–June 2018 exhibition based on my research. *The Codex and Crafts in Late Antiquity* is the catalogue written to accompany the exhibition.

Among the scholars on whose work I based much of my research and whose names are included in the bibliography, I need to specially mention the scholarly work of Theodore C. Petersen on the Coptic bindings of the Morgan Library and Museum, most of which still remains unpublished. Petersen's analytical approach, his detailed accounts, and his extremely helpful drawings have inspired and guided my research in many different ways.

The Codex and Crafts in Late Antiquity would not have been at all possible without the generosity and creative thinking that characterize Bard Graduate Center. I am most grateful to all the staff, but especially to Dean Peter Miller; Ivan Gaskell, professor of cultural history and museum studies and head of the Focus Projects; Marianne Lamonaca, chief curator and associate director of the Gallery; Caroline Hannah, associate curator; Elena Pinto Simon, former dean for academic administration, students, and alumni affairs; Eric Edler, registrar and associate preparator; and Jesse Merandy, director of the Digital Media Lab. Special thanks are due to the team who worked toward making this publication possible: Daniel Lee, director of publishing; Jocelyn Lau, junior designer; Kate Dewitt, art director; and Alexis Mucha, manager of rights and reproductions. The students of my class in the fall semester 2016—Jaime Ding, Emily Field, Julia Lillie, Aleena Malik, Sarah Reetz, and Darienne Turner (who has also been my research assistant)—have helped enrich and clarify various issues, including the iconographical evidence and the digital component of the exhibition. My thanks also go to exhibition designer Jeremy Johnston of Darling Green.

I am extremely grateful to the staff at the Morgan Library and Museum: Roger S. Wieck; Melvin R. Seiden, curator and department head; William M. Voelkle, senior research curator; John Vincler, head of reader services; Maria Molestina-Kurlat and Sylvie L. Merian, reader services librarians; Maria Fredericks, Drue Heinz Book Conservator; and Frank Trujillo, associate book conservator. They all made my days at the Morgan pleasant, extremely informative, and memorable. Needless to say, without the contribution of the iconic bindings from the Morgan collection, the Focus Gallery exhibition would not have been possible at all. Also in Manhattan, in addition to the excellent library at BGC, two other libraries fed my curiosity: the Butler Library at Columbia University and the Watson Library at the Metropolitan Museum of Art. To the people working in all three, I am extremely thankful.

I feel equally grateful to the staff at the Chester Beatty Library in Dublin: Jessica Baldwin, head of collections and conservation; Kristine Rose Beers, senior conservator; Julia Poirier, conservator; and Hyder Abbas, assistant librarian. I also offer thanks to Don Federico Gallo, the director of the Ambrosiana Library in Milan, and Stefano Serventi, who is responsible for the cataloguing of manuscripts of the library, for allowing me to examine precious manuscripts in their collections. I would never have been able to complete this study without the extraordinary experience of working in the library of St. Catherine's Monastery at the foot of Mount Sinai in Egypt, both for my PhD research and for the assessment of the collection in the context of the St. Catherine's Library Conservation Project. The very questions that were the basis for this study were born while handling and closely examining the unparalleled collection of early codices in the monastery's library. To Archbishop Damianos and all the fathers of the monastery, especially librarians Father Simeon and Father Justin, I am deeply grateful. The same goes for Nicholas Pickwoad, professor and project leader of the St. Catherine's Monastery Library Project based at the University of the Arts London, where he is director of the Ligatus Research Centre.

I am indebted to the curators of the various collections that I consulted for spending some of their precious time with me, letting me see up close many precious artifacts, some of which they lent for the exhibition. From the Metropolitan Museum of Art: Helen Evans, Mary and Michael Jaharis Curator for Byzantine Art; Sheila Canby, Patti Cadby Birch Curator in Charge, Department of Islamic Art; and textile conservator Kathrin Colburn; as well as Eva DeAngelis-Glasser, assistant for administration; and Isabel Kim, senior collections management associate from the Antonio Ratti Textile Center. I am also indebted to Edward Bleiberg, curator of Egyptian, classical, and ancient Near Eastern art, Brooklyn Museum; John S. Hansen, collections manager in the Division of Anthropology, American Museum of Natural History; Lisa R. Brody, associate curator of ancient art, Yale University Art Gallery; and Mina Moraitou, curator at the Benaki Museum in Athens.

A number of others have also shared their knowledge and expertise with me during my research. I am most grateful to Joy Boutrup, associate professor emerita at Kolding School of Design, who has been extremely helpful and patient with me, especially in the early stages of this research when I was struggling to grasp the world of textiles. I would like to thank Frances Pritchard, curator of textiles at the Whitworth Gallery of Manchester University; and Willeke Wendrich, basketry expert and director of the Cotsen Institute of Archaeology at UCLA. Help on some of the most unglamorous but very important items for my research arrived through Regina de Giovanni, who was most generous and willing to discuss and dedicate her time and skills replicating socks from late antique burials in Egypt made with the cross-knit looping technique. Given that, except for a couple of fragments, it was not possible to identify original examples in any American collection, her replicas were all the more appreciated in the context of the exhibition and catalogue. Ursula Mitra has been helpful and generous in different ways, lending us some of her facsimiles of historic bookbindings for the exhibition. The illustrations throughout this book include many drawings that appear as supplements to verbal descriptions of complex techniques; unless otherwise noted, the drawings are by the author.

I would also like to thank my friends Maria Georgaki, Anna Gialdini, Lazaros Apokatanidis, Efrosini Theou, Stamatis Zoumbourtikoudis, Ilias Tsolakopoulos, Athanasios Velios, Nikolas Sarris, Ioannis Melianos, and Christina Margariti. Last but not least, the anonymous reviewers deserve to be thanked for their important and valuable comments, as well as the copyeditor, Carolyn Brown, and the proofreader, Christine Gever. They all helped improve the book, although the responsibility for any mistakes is solely mine.

Introduction

The Innovation of the Codex in Late Antiquity

The most momentous development in the history of the book until the invention of printing was the replacement of the roll by the codex.

—Roberts and Skeat, *The Birth of the Codex*

T he codex appears to have been a Roman innovation, one of the latest examples of Roman ingenuity and certainly one of the most important for the history of civilization. Throughout Greco-Roman antiquity, the standard format for an extended written text had been the papyrus roll. Literary evidence suggests that the Romans, following the structural and functional principles of the wooden tablet codex, turned from wooden "leaves" to papyrus and parchment—already used for informal notebooks—thus producing the book as we know it today.[1]

Among the Romans, a *caudex* was a number of wooden tablets fastened together, also called *tabulae*. Literary evidence indicates that tabulae coexisted with *membranae*, informal notebooks made with leaves of parchment (*membrana*) connected together.[2] In wall paintings from Pompeii created before the eruption of Vesuvius in AD 79, we can identify, among the writing implements and books represented, various book formats used at the time—wooden tablets with one or more "leaves," waxed or whitewashed; papyrus rolls rolled or open with a title tag (*sillybus*) attached; an open *capsa* (container for rolls); various inkpots and writing pens; and even a scraper-like tool, presumably for leveling the wax of the tablets—but nothing that could be even tentatively described as a parchment or papyrus notebook or codex (fig. 1).[3] Two centuries later, in a wall painting in the tomb of Trebius Iustus, a young man who died at the age of twenty-one, is represented with a number of different book formats around him, including a capsa with some rolls inside, thirteen tablets that probably constitute a codex with a long strap extending from one of its corners, and a single board with what appear to be three holes on each of the short sides. He is holding a papyrus or parchment codex, which can be clearly distinguished from the wax tablet codices that lie open beside him (fig. 2).

Although codices are certainly books, books did not come exclusively in the form of codices, especially in the Greco-Roman period, when the standard form of the book was the roll. Iulius Paulus, praetorian prefect between AD 228 and 235, clearly states in his *Sententiae*: "When books are bequeathed, rolls of papyrus, or of parchment and wood-slabs, are included, and codices, as well. By the designation books, not merely rolls of papyrus, but also any kind of writing which is contained in anything, is understood."[4]

Fig. 1 Still life with writing implements, four-leaved wax tablet codex, spatula (possibly for erasing), double inkwell with reed pen (*calamus*), and papyrus scroll. Fresco from the House of Fabius Secundus, Pompeii, before 79 BC. Museo Archeologico Nazionale, Naples, inv. no. 4676.

Fig. 2 Wall painting representing twenty-one-year-old Trebius Iustus, among different book formats, stylus pockets, and a *capsa*, late 3rd–4th century AD. Tomb of Trebius Iustus, Via Latina, Rome.

Despite the predominance of Latin literary evidence from Rome about the innovation and use of the early codex, the earliest physical evidence we have is neither Roman nor Latin but rather from Greek texts unearthed in Egypt. The apparently new book format appears initially to have had little success for Latin and Greek literary texts, for which the roll was used exclusively. It appears to have been wholeheartedly accepted by Christians, however, so much so that the new religion, which quickly spread from the Middle East to the rest of the Mediterranean and Western Europe (fig. 3), is often credited with establishing the codex as the standard format of the book, gradually supplanting the roll. The change took place between the second and fifth centuries AD for reasons and under conditions that have been debated by many scholars for many years. Broadly speaking, the reasons proposed for the special preference for the codex among Christians are utilitarian, social, and religious. Utilitarian reasons include the various practical advantages of the codex, such as the ease of reference, the capacity to accommodate longer texts (because both sides of a leaf could be used), the arguably lower cost, and easier portability. Social reasons include the social status of Christians, their education, their need to integrate into the Roman society of Egypt and at the same time distinguish themselves from the Jews (who were using rolls for their sacred books), and possibly their unfamiliarity with the roll book format.

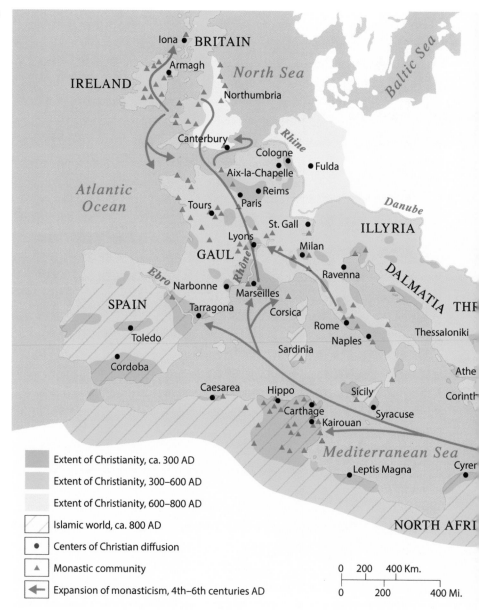

Fig. 3 Spread of Christianity, ca. AD 300–800, and major monastic communities, 4th–6th century AD.

Iona • BRITAIN

Armagh •

IRELAND

North Sea

Northumbria

Canterbury •

Rhine

Cologne •

• Fulda

Aix-la-Chapelle •

Atlantic Ocean

Reims •

Tours • Paris •

St. Gall •

Danube

ILLYRIA

Lyons •

Rhône

Milan •

GAUL

Ebro

Ravenna •

DALMATIA THR

Narbonne •

Marseilles •

SPAIN

Tarragona •

Corsica

Rome •

Thessaloniki

Toledo •

Sardinia

Naples •

Cordoba •

Athe

Caesarea •

Hippo •

Sicily

Corinth

Carthage •

Syracuse •

Kairouan •

Mediterranean Sea

Cyrer

Leptis Magna •

NORTH AFRI

Extent of Christianity, ca. 300 AD

Extent of Christianity, 300–600 AD

Extent of Christianity, 600–800 AD

Islamic world, ca. 800 AD

• Centers of Christian diffusion

▲ Monastic community

← Expansion of monasticism, 4th–6th centuries AD

| 0 | 200 | 400 Km. |
| 0 | 200 | 400 Mi. |

Detail of Egypt

St Menas Monastery • Alexandria
St Macarius Monastery

NITRIA DESERT
Kellia *SINAI*
Scetis *PENINSULA*
Fayum • Cairo
Hamuli St Anthony Monastery
• Naqlun St Catherine's Monastery
Archangel Michael Monastery St Paul Monastery

Antinoe •

St Apollo Monastery, Bawit

Nile R.

Sohag • Akhmin-Panopolis
Red and White Monasteries • Tabennisi
Nag Hammadi • Thebes
Dakhleh Oasis • Esna
• Edfu

St Simeon Monastery • Asuan

Red Sea

LOWER NUBIA

Faras • • Qasr Ibrim

Dnieper

Black Sea

Sinope •

Constantinople

GEORGIA

ARMENIA

ANATOLIA Edessa

Tigris

hesus

Antioch • *Euphrates*

Dura-Europos

Cyprus SYRIA
• Damascus
Caesarea

PALESTINE Jerusalem

exandria

SINAI ARABIA

EGYPT

Nile R.

Red Sea

Finally, religious reasons include specific texts—Mark's Gospel, the Four Gospels, or the letters of Paul—that may have been compiled in codex format either originally or very early on and might have acted as venerated, and subsequently imitated, prototypes.[5] None of the various theories has been widely accepted. Some are less convincing than others, and some have been retracted by the scholars who proposed them. Nevertheless, one hypothesis does not necessarily exclude the others, and there is no reason why all factors—utilitarian, social, and religious—should not be treated as simultaneously valid and operative. What is certain is that the change from roll to codex took place in the early centuries of late antiquity and that the codex should be understood not as an invention but as an innovation—that is, as the result of a "process of bringing new methods, ideas or practices to an existing technology, which substantially modifies an existing technology in a given society."[6]

The usual sequence in any new addition to the human array of artifacts is invention or innovation, followed by diffusion and use, usually long after the initial invention or innovation.[7] In other words, it is perfectly logical—and in fact a common process—for the innovation of the codex to have taken a couple of centuries to become established and widely used, as surviving artifacts suggest. It is also true that the spread of the same innovation can vary considerably among different contemporary cultural settings.[8] Thus, that Christians apparently adopted the codex straight away, while the pagans may not have adopted it until a couple of centuries later, is not strange in itself, but it does require explanation. In technological terms, the codex can be understood as the result of a long evolutionary process from simpler, well-established, widespread technologies—such as rolls, papyrus and parchment notebooks, and wooden tablets—to the more complex multigathering codex. The codex can be also understood as a synthesis of different techniques, borrowed or transferred from such different crafts as shoe and sock making, woodworking, and fabric making (including textiles, mats, and basketry).[9]

Where did the codex innovation take place? Almost all the earliest codices to have been preserved—not as a collection of loose leaves but as entire bound books—have been unearthed in Egypt because the dry climate helped their preservation. As a result, there is a long-established and deep-rooted notion that the making of the codex, at least in its structural components, owes much to the Copts (the native Egyptian Christians). This is a thesis, however, that needs to be reconsidered.[10] Egypt has definitely played a central role in the history of the book—if not for anything other than its preservation of all the existing physical evidence from the earliest centuries. Nevertheless, this role should not be overstated. Iconographical evidence from the late Roman and late antique periods is enough to prove that books around the Mediterranean were made in the same way, or at least looked the same, following what could be described as a bookbinding lingua franca. There is simply not enough evidence to identify where in the Mediterranean this common language might have originated, but as will be shown, the available evidence points to a Roman cultural and technological provenance for the codex. If the geographical origin is to be identified with Roman Egypt, this still needs to be convincingly proved. Otherwise, it would be like attributing the composition of the *Shepherd of Hermas* to Egypt because the earliest physical copies were all found in Egypt, when it is known that the text was originally composed in Rome.[11] It is also perhaps misguided to look for a specific place of origin for the codex in the first place. As will be shown, the codex was not invented overnight but was rather the result of a long evolutionary process that may have taken place simultaneously in various places around the Mediterranean—Rome, Alexandria, Jerusalem, Antioch, Constantinople, Caesarea, and various Eastern Mediterranean monastic and ecclesiastic communities. Which of these might have exercised the decisive role is still to be established, but I think that we should look to the large urban and cultural centers of late antiquity, keeping in mind that although we have some examples of the highest standards of manuscript production from the fourth century AD, such as the Codex Sinaiticus, we lack evidence of the corresponding highest standards—not only aesthetic but technical—of bookbinding from the same period.[12] The establishment of the codex in the first five centuries AD should be viewed not as an epiphany but as a gradual, complex, nonlinear evolutionary process. Out of the various manifestations of the book, used simultaneously and for different purposes in the same time and place, the format with the greatest number of advantages, not only practical and economic but also symbolic and social, must have been selected and adopted. An example of the circulation of texts in several different formats is provided by the three sisters of Thessaloniki—Agape, Irene, and Chione—who in AD 304 were arrested and ultimately executed because they were found to possess a number of Christian writings on parchments (*difteras*), books (*biblia*), wooden tablets (*pinakidas*), codices (*kodikelous*), and pages (*selidas*).[13] The canonization of the Gospels followed a similar process over the course of a couple of centuries. Kim Haines-Eitzen explains that, "among the some 5,400 Greek manuscripts of the New Testament texts, for example, no two are identical; more relevant, perhaps, is the fact that some fifty-two extant manuscripts that can be dated to the period from the second century to the fourth exhibit more differences and variations than the thousands of later manuscripts."[14] If, despite the intense research on the subject and the substantial literature produced, there is no consensus about the process of writing and canonizing the Gospels between the first and third centuries,[15] then there is not much hope of defining with accuracy the process by which the codex's structure was established. Based on the fragmentary available evidence—physical, literary, and iconographical—and the instinctive human tendency to fill in the gaps to make sense out of the facts, we should always be aware of the danger of a "fantasy reconstruction."[16]

How did codex innovation spread? Although specific parts of the bookbinding process require hands-on instruction—for example, the sewing of the codex or the endbands—it is not difficult to imagine that, up to a point, bookbinding techniques in those early days could have been transferred along with the actual bound books, even if not through the person who actually bound them.[17] A bookbinder or an artisan would not have had much difficulty deducing or re-creating all or parts of the process, possibly even by partly undoing a bound codex to view its internal features, such as how the gatherings were sewn.

The initial phases of the evolutionary process—represented by the wooden tablet and single-gathering codices—that eventually led to the multigathering codex possibly took place in a secular rather than a Christian context. From the fourth century onward, however, as Christianity was gradually established as the official religion of the Byzantine Empire and as monasticism spread, the context of book production appears to have changed. Although private copying of books would not have stopped and books would have continued to be copied in secular settings,[18] most book production occurred in monastic or ecclesiastical settings. For example, when Constantine wanted fifty copies of the Bible for the churches of the new capital city of the empire, Constantinople, he ordered them from Eusebius, bishop of Caesarea Maritima in Palestine, in AD 332.[19] Chrysi Kotsifou notes that "in late antiquity, centers of book production were primarily if not exclusively in monasteries."[20] And as will be argued in the final chapter, monasteries contained all the necessary technical skills and knowledge, as well as the craftsmen, to turn a number of written leaves into a bound codex.

I n antiquity, the word "codex" was used to designate a set of wooden tablets fastened together and used for both formal and informal writings. The word has evolved to reference a handwritten and handbound book with leaves arranged in gatherings that are sewn or otherwise fixed together to preserve their correct sequence and allow for safe handling and preservation of the text: the typical format of a book as we know it today. The fundamental features of a codex are multiple individual leaves with text upon both sides, permanently connected together along one side through a separate element, whether thread, cord, or leather.[21] In *The Codex and Crafts of Late Antiquity*, we will look at codices on the basis of structure and format.

We will consider three different types of codex: the wooden tablet, the single-gathering, and the multigathering codex. Wooden tablets consist of a number of relatively thin rectangles of wood connected together, while single and multigathering codices are made of papyrus or parchment folios—single folios either stacked one on top of the other or folded in the middle to form a gathering, or quire.[22] In a multigathering codex more than one gathering is sewn to another (figs. 4 and 5). Archaeological, literary, and iconographical evidence all suggest that the three types coexisted for

Fig. 4 Anatomy of a multigathering codex as seen from spine. The codex represented here is an idealized example combining features found in various manuscripts bound during the 8th– 10th century AD in the Eastern Mediterranean.

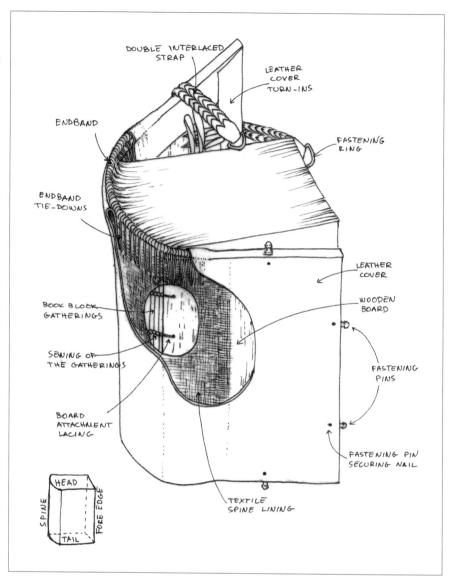

DOUBLE INTERLACED STRAP

LEATHER COVER TURN-INS

ENDBAND

FASTENING RING

ENDBAND TIE-DOWNS

LEATHER COVER

WOODEN BOARD

BOOK BLOCK GATHERINGS

SEWING OF THE GATHERINGS

FASTENING PINS

BOARD ATTACHMENT LACING

FASTENING PIN SECURING NAIL

TEXTILE SPINE LINING

HEAD

SPINE

FORE EDGE

TAIL

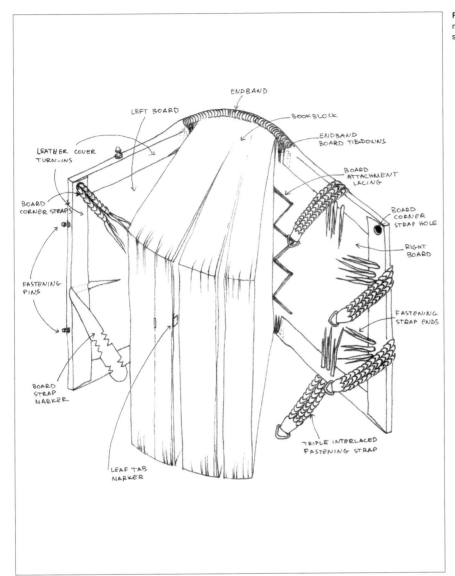

Fig. 5 Anatomy of a multigathering codex as seen from fore edge.

ENDBAND

LEFT BOARD

BOOKBLOCK

ENDBAND
BOARD TIE-DOWNS

LEATHER COVER
TURN-INS

BOARD
ATTACHMENT
LACING

BOARD
CORNER STRAPS

BOARD
CORNER
STRAP HOLE

RIGHT
BOARD

FASTENING
PINS

FASTENING
STRAP ENDS

BOARD
STRAP
MARKER

TRIPLE INTERLACED
FASTENING STRAP

LEAF TAB
MARKER

centuries alongside the roll, although ultimately the multigathering codex supplanted the others and has remained the primary form of the book in the Western world to the present.

Bookbinding is both the process of turning a number of written folios into a functional book as well as the actual physical container of the written or printed leaves of a book—also called a book block. Although the binding of a book is usually understood as its decorated cover, it also includes features that are not necessarily visible—for example, the sewing of the gatherings and the connection of the boards to the book block—but which make up the structure of the codex and which can be studied in the same way as any other artifact.

The history of the codex, especially the early codex, has been until recently a rather two-dimensional narrative, focusing on the page rather than on the three-dimensional book. The very notion that a codex would have been bound in some way is usually overlooked. Like a child born to illustrious parents, bookbinding has been overshadowed by the study of the texts enclosed within the binding. More probably, though, it was and still is difficult for textual scholars to grasp the craftsmanship—its peculiarities, characteristics, necessities, and consequences—required for turning a number of written leaves into a functional, three-dimensional object—what we call (or at least used to call until the appearance of e-books) a book. This has had two consequences: first, it allowed codices unearthed in good condition to be disbound, often carelessly, thus depriving them of important features and information; second, it excluded from scholarship important evidence that might be useful for answering persisting questions about the appearance and establishment of the codex as the standard book format since late antiquity.

In 1958 Berthe van Regemorter wrote: "The texts of the old rolls or volumina have been studied by many scholars. So have the texts of the old codices and their paleography, but it is only comparatively recently that it has been realized that the general appearance of those codices, the way their gatherings are put together, the way they are bound, are of great importance also. Every part of a codex must be examined for clues to its provenance and to be able to confirm or disprove the date which the paleographer has ascribed to it."[23] Even a scholar of the magnitude of Eric G. Turner, who allegedly "knew vastly more early codices than . . . practically anyone else before or after,"[24] apparently failed to make a distinction between supported and unsupported sewing.[25] Supported sewing was typical of the books bound in Western Europe from as early as the eighth or ninth century, while unsupported sewing was employed in all the books bound in the Eastern Mediterranean tradition from the earliest times until the seventeenth and eighteenth centuries, and in some cases until the present. Despite a steady production of scholarship on the early history of the codex, its binding is still mostly left out of consideration. It is also indicative that most scholars writing on the subject seem to be unaware of one of the soundest and clearest books on the history of bookbinding ever written, János A. Szirmai's *The Archaeology of Medieval Bookbinding*, which would have at least provided the basic information on a subject understandably unfamiliar to textual scholars.

The aim of *The Codex and Crafts of Late Antiquity* is to offer an interdisciplinary approach to the history of the making of the codex, an account not of why it supplanted the roll as the standard form for extended written texts but rather of how it did so. With a focus on the artifactual aspects of the codex and its making, this study traces and describes the different techniques involved, whenever possible considering them in parallel with other more or less similar techniques that have been used by the same people, in the same time, and in the same cultural and geographical context in different crafts, such as weaving and other fabric-making techniques, basketry and mat making, leatherworks, and shoemaking. This book and the exhibition it accompanies trace these interwoven strands. In the preface to his *Books and Readers in the Early Church*, Harry Y. Gamble points out that "it is always a risk to cross disciplinary boundaries (there is a bewildering landscape on the other side) but it is often a risk worth taking."[26]

The idea that the techniques involved in making the codex were borrowed and adapted from other crafts is not new. But so far, although other scholars have alluded to it, there has been no consistent research on the subject. For some techniques, the relationships among crafts are so straightforward and clear that they have been noted repeatedly. In 1910 Theodor Gottlieb noted the connection between the decoration of the leather covers of bookbindings and the decoration of shoes, a connection that Paul Needham also stresses in relation to Morgan codex M.569 (discussed in chap. 8).[27] Theodore C. Petersen notes that the same decorating tools and techniques that were used for bindings were also used for "girdles, aprons, sandals" and "beautifully decorated shoes, boots and slippers found in the necropolis of Achmin," and suggests that the joining of codex gatherings must have been based on various techniques, such as sewing, lacing, and braiding, used when the codex first appeared.[28] Gary Frost clearly states that "the resources of many crafts must have been assimilated into early codex bookbinding. . . . Crafts of sewing leather tents and containers would also be relevant."[29] For other features of the codex, however, the relationship to other crafts is not immediately clear, although I think that once a more open-minded and wider view is adopted, it is quite easy to see not just a vague affinity between books and other artifacts but often specific techniques and processes that they share.

The pages that follow offer a technical history of the codex from the first centuries AD to roughly the tenth century and try to illuminate the techniques used to make the different types of codices in the general technical context of late antiquity. In terms of the multigathering codex, I argue that its structure evolved from a combination of the wooden tablet codex and the single-gathering codex; both, structurally, are rather simple book formats used extensively in Greco-Roman antiquity.

The transition from roll to codex took place in late antiquity, which generally speaking is the period between the third and the eighth centuries AD.[30] According to the geographical and cultural context to which it refers, the term "late antiquity" encompasses a number of more or less distinct historical and cultural periods—early Christian, Byzantine, late Roman, and early Islamic—to mention those that are most relevant to our discussion. The term also encompasses the major monotheistic religions—Judaism, Christianity, and Islam—with all their variations and interactions around the Mediterranean in different arenas, from the religious and political to the cultural, technological, and art historical. Although the upper boundary of late antiquity is never defined as later than the eighth century, for practical reasons I will also be including material from the ninth and tenth centuries. If we were to exclude the bound codices of the ninth and tenth centuries that have survived in a more or less complete state (that is, including their bindings), we would be left with only a handful of examples from the fourth through eighth centuries. There are, however, a substantial number of ninth- and tenth-century codices preserved in European and American libraries as well as in their original monastic or religious settings, among which are the St. Catherine's Monastery library in Sinai, the Kairouan Mosque in Tunis, and the Great Mosque at Sana'a in Yemen. Discoveries such as those from Kairouan and Sana'a have made it possible to establish that until the tenth and eleventh centuries, the bindings of Christian and Islamic books still preserved many technical and decorative elements in common, most of which were to disappear in later centuries, to an extent concealing their common past.[31]

Available evidence indicates that changes in the way books were bound and decorated in the Eastern Mediterranean occurred very slowly. Techniques used in the earliest preserved codices survived virtually unchanged in the Byzantine tradition from the fifth century to the seventeenth century; in the Islamic tradition, into the eighteenth and nineteenth centuries; and in the Ethiopic tradition, possibly to the present. This is remarkable for various reasons and is rather telling in terms of the characteristic conservatism of bookmaking in the Eastern Mediterranean, which can partly be explained by the book's association with and prominent role in religion. In contrast, Western European bookmaking techniques, which originally followed the same technical principles as the Eastern Mediterranean bookbinding traditions, had by the eighth century developed substantial technical changes. These changes were so marked that Western European bindings are treated as a separate evolutionary branch in the history of bookbinding. Through the invention of printing and the circulation of books printed and bound in the West, Western bookbinding techniques eventually supplanted those that had been used in the Eastern Mediterranean since the early Christian centuries.

The evidence for reconstructing the history of the Eastern Mediterranean tradition can be separated into three types: physical, iconographical, and literary.[32] Inevitably, evidence for each is fragmentary, especially the first, which is our primary source of information; both iconographical and literary evidence, however, can at times be surprisingly accurate and helpful. Among the three, iconographical evidence has attracted the least attention in the study of the early codex, even though such evidence is standard in other areas of research, such as art history, archaeology, and material culture.[33] Iconographical evidence, if properly and consistently studied along with physical evidence, can help fill the many gaps related to the format and making of the book in late antiquity.[34] This kind of evidence probably represents our only chance of reaching an understanding of the physical codex of late antiquity outside Egypt (where most of the physical evidence has been preserved because of favorable climate conditions). The main problem with exploiting iconographical evidence is that unless we have related physical evidence to which to compare it, we are often left with an image that we cannot quite decipher. Nevertheless, even when there are features we do not understand, they are often consistently represented in different works of art that are not connected in any apparent way, which suggests that they reflect the reality of the codex at the time, even if that reality remains unclear.

The individual chapters dealing with the three different types of codex differ greatly in length. Part II, dedicated to the multigathering codex, constitutes roughly two thirds of the text primarily because this type of codex is much more complex than the wooden tablet and single-gathering codices. The separate components of a multigathering codex combine different techniques, each of which requires separate consideration. Given the number of surviving multigathering codices from late antiquity and the variations among them, there is far more information to consider than for either the wooden tablet or the single-gathering codex.

The Codex and Crafts in Late Antiquity is in no way an exhaustive treatment of the subject. Rather, it should be regarded as an introduction to a different way of looking at books, offered in the hope that more research will correct possible distortions and misinterpretations and will sharpen our understanding of this most intriguing artifact.

Notes

1. For the literary evidence, see Roberts and Skeat, *The Birth of the Codex*, 15–34.
2. See ibid., 15–23.
3. On the wall paintings and the various implements depicted, see Elizabeth Meyer, "Writing Paraphernalia, Tablets and Muses." According to Meyer, "in four different types of paintings—still lifes, literary vignettes, mythological paintings, and 'portraits'—writing implements, and especially wooden tablets, appear, but only in the first type of painting do the objects correspond well to the objects themselves. These still lifes aim at representing the real; the others do not" (589). Nevertheless, I would contend that the representation of objects such as wooden tablet codices are also usually remarkably accurate in the other types of painting.
4. *Sententiae* 3.6.87: "Libris legatis tam chartae volumina vel membranae et philyrae continentur: codices quoque debentur: librorum enim appellatione non volumina chartarum, sed scripturae modus qui certo fine concluditur aestimatur." Quoted in Roberts and Skeat, *Birth of the Codex*, 32. Translation based on http://ccat.sas.upenn.edu/rak /courses/735/book/codex-rev1.html.
5. Literature on the subject of the early codex and its relation to Christianity is extensive. See, for instance, Roberts and Skeat, *Birth of the Codex*; Hurtado, *The Earliest Christian Artifacts*; Gamble, *Books and Readers*; Millard, *Reading and Writing in the Time of Jesus*; Stanton, *Jesus and Gospel*; Meyer, "Roman Tabulae, Egyptian Christians, and the Adoption of the Codex."
6. Hollenback and Schiffer, "Technology and Material Life," 322.
7. See Edgerton, "From Innovation to Use," 116. A similar example of innovation is the production of glass, which at least in the Near East became much more common in late antiquity than it had been before. Although most of the techniques may have been known, their use became much more widespread and certainly more refined, producing some great masterpieces, for example, in gold glass and cage cups. See Lavan, "Explaining Technological Change," xxv, xxvii.
8. On the spread of innovation in different cultural settings, see Edgerton, "From Innovation to Use," 116.
9. For a discussion of the term "fabric" and the different artifacts it includes, see Emery, *The Primary Structure of Fabrics*, 208–210; Wendrich, *The World according to Basketry*, 31–37.
10. See Boudalis, "The Bindings of the Early Christian Codex."
11. For the fast spread of the *Shepherd of Hermas* from Rome to Egypt, see Hurtado, *The Earliest Christian Artifacts*, 27.
12. "Codex Sinaiticus" refers to a fourth-century codex of the Old and New Testament, for the most part preserved in the British Library (Add. MS 43725), with small parts of it distributed in other collections, including St. Catherine's Monastery in Sinai, Egypt, where the manuscript had been preserved until the middle of the nineteenth century. Other codices that are still preserved in the library of St. Catherine's Monastery will be cited in this book as "Sinai codex," with a language and reference number.
13. For an account of the three sisters, see Musurillo, "The Martyrdom of Agape, Irene and Chione," 286–287.
14. Haines-Eitzen, *Guardians of Letters*, 106.
15. Olson, "The Sacred Book," 1:13–14.
16. See Greene, "Historiography and Theoretical Approaches," 73.
17. On the ease of traveling and communication in the period, see Hurtado, *The Earliest Christian Artifacts*, 26–27; Gamble, *Books and Readers*, 96, 142–143.
18. See Cavallo, *Libri editori e publico nel mondo antico*, 91–92.
19. See Eusebius, *Vita Constantini*, 4.36–37. See also Gamble, *Books and Readers*, 79.
20. Kotsifou, "Books and Book Production," 50. See also Gamble, *Books and Readers*, 121–122.
21. Very simple structures such as diptychs (two wooden tablets connected together) do share the same basic technical and structural features and could thus be treated as protocodices.
22. A "quire" (from the Latin *quaternio*, meaning "group of four") is a set of four sheets of

papyrus, parchment, or paper folded in the middle to create eight leaves, or sixteen pages. The term has commonly been used to signify a gathering independent of the number of folded leaves and is synonymous with the word "gathering." See Turner, *The Typology of the Early Codex*, 55.

23. Regemorter, "Some Early Bindings from Egypt in the Chester Beatty Library," 152. The exclusion of bookbinding from the "official" history of the book is very common. In Eliot and Rose's *A Companion to the History of the Book*, for instance, there is not even a small chapter dedicated to bookbinding, which is only treated very briefly in passing.

24. Bagnal, *Early Christian Books in Egypt*, 12.

25. Turner's description of the binding process of the multigathering codex is rather problematic, reflecting a poor understanding of the method. He states that in a multigathering codex, "each set of folded sheets is 'gathered' together and stitched, i.e., the sets form 'gatherings.' One set of threads holds each gathering together. If the book has a binding, a second set passes horizontally through the first set and unites the gatherings; it is taken then through them across the spine of the book and secured to the front and back binding covers" (*Typology of the Early Codex*, 55).

26. See Gamble, *Books and Readers in the Early Church*, xi.

27. Gottlieb, *Bucheinbände*, cited in Arnold and Grohman, *The Islamic Book*, 35, note 159; Needham, *Twelve Centuries of Bookbindings*, 15.

28. See Petersen, "Coptic Bookbindings in the Pierpont Morgan Library," 25, 80–82.

29. See Frost, "Adoption of the Codex."

30. On late antiquity, see, for instance, Brown, *The World of Late Antiquity*; Cameron, *The Mediterranean World in Late Antiquity AD 395–600*; Bowersock, Brown, and Grabar, *Late Antiquity*.

31. Of course, to judge from the surviving material, there also seem to be major differences: for example, the Islamic box bindings in the ninth and tenth centuries and the fore-edge flap, ubiquitous in Islamic bindings after the twelfth century. On these features, see Petersen, "Early Islamic Bookbindings and Their Coptic Relations"; Di Bella "An Attempt at a Reconstruction of Early Islamic Bookbinding."

32. On the combination of different types of evidence, see Lavan, Swift, and Putzeys, *Objects in Context, Objects in Use*, esp. the introductory notes, 3–27.

33. For an example of the use of iconographical evidence in material culture studies and archaeology, see Parani, *Reconstructing the Reality of Images*. For a general consideration of the subject, see Burke, *Eyewitnessing*, esp. 81–102. See also Haskell, *History and Its Images*, esp. pt. 2; Lavan, Swift, and Putzeys, *Objects in Context, Objects in Use*, 11–13.

34. On the possibilities and limitations of the use of iconographical evidence to clarify the construction, appearance, and use of the early codices, see Boudalis, "Clarifying the Structure, Appearance and Use of the Early Codex around the Mediterranean Basin."

Part I

The Precursors of the Multigathering Codex

One

The Wooden Tablet Codex

Among the ancients a structure formed by joining boards together was called caudex, whence also the tablets of the law are also called codices.

—Seneca, *De brevitate vitae*, AD 49

T his chapter examines the different types of wooden tablet codex used in antiquity until the fifth and sixth centuries AD, particularly methods for connecting individual wooden tablets within the codices that may have been incorporated into the structure of the multigathering codex. Several wooden tablet codices have been preserved, although as Rafaella Cribiore points out, "the provenance of less than a quarter of tablets is known, and not only is this percentage too low to allow any conclusions but apparently only a few locations (e.g., Antinopolis) are represented."[1] Elizabeth Meyer counts fifty-nine examples, divided among school, account, and legal tablets, that have been preserved from the first two centuries of the Christian era; there are many more from the following centuries.[2]

We tend to think of wooden tablet codices as rather cumbersome and inelegant, but there are actually examples that are both functional and elegant, made of thin and finely worked slabs of wood. The belief that the wooden tablet codex is the predecessor of the multigathering codex is common among scholars[3] and is based on the similarities of their functional aspects, despite the difference in the material of the leaves (wood as opposed to papyrus or parchment and, later, paper). Both have a rectangular format with leaves—both sides of which can have writing—that are connected along one of the long sides permanently yet flexibly, so that they can open and close while safely preserving the order of the text leaves.

The word "codex" derives from the Latin *caudex*, meaning "block of wood." Evidence of the long history of wooden tablets in the Eastern Mediterranean can be found in the Uluburun diptych, which dates back to the fourteenth century BC (fig. 6).[4] Iconographical, literary, and archaeological evidence indicates that tablets of one type or another were used for writing among the Hittites as early as the eighth century BC and are also mentioned by Homer.[5] Representations of wooden tablets on Greek vases (fig. 7), wall paintings (see figs. 1 and 2, introduction), and even sculpture[6] suggest how widespread their use was in antiquity. There was not a single type of wooden tablet codex but rather several, differing not only in their makeup but also in the type of text they contained. The tablets could be

Fig. 6 Representation of the Uluburun diptych. Drawing by Netia Piercy. From Payton, "The Ulu Burun Writing-Board Set," 102, fig. 2.

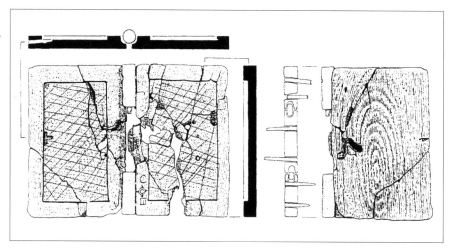

Fig. 7 Kylix depicting schoolboy with wooden tablets. Attributed to the painter of Munich 2660. Kylix, ca. 460 BC. Terracotta (earthenware), red-figure technique. The Metropolitan Museum of Art, Rogers Fund, 1917, 17.230.10. Cat. 6.

plain, with possibly only a coat of gum Arabic or whitewash (in which case the tablet was called an *album* or *leucoma*) to facilitate writing with pen and ink and to prevent the ink from being absorbed by the wood.[7] Alternatively, and more commonly, the tablet would be slightly recessed to contain wax, possibly mixed with carbon pigment or other colors when melted,[8] so that letters could be incised on it with a stylus and then leveled later to erase the writing.[9] Such tablets could be small enough to fit in the palm of a hand—in which case the Romans called them *pugillares* (fig. 8)—or they could be much bigger.[10]

The edges of the tablets could be cut straight, or the upper edge could be notched or even have tabs extending from the fore edge (opposite the sewn edge) for facilitating handling and turning individual "pages."[11] The codices

Fig. 8 Codex composed of five tablets containing writing exercises, 4th century AD. Wood and wax. Brooklyn Museum, Charles Edwin Wilbour Fund, CUR.37.1908E, 37.1909E, 37.1910E, 37.474E, 37.473E. Cat. 1, see also cat. 2.

could contain two tablets (a diptych), three (a triptych), or multiple tablets (poliptychs) (fig. 9; see also figs. 6–8, figs. 1 and 2, introduction).[12] Martial refers to wooden tablet codices made out of precious citron wood and ivory, as well as codices with three and five tablets.[13] They could also have two holes on the fore edge, one on each of the upper and lower covers or boards, presumably for fastening laces (fig. 10). There were usually small squares in the middle of the tablets so that the wax would not stick adjacent "pages" together when the codex was closed (fig. 11; see also fig. 1, introduction).[14]

Tablets can generally be distinguished by content. There were school tablets, containing school exercises; account tablets for various types of financial records; and legal tablets for such documents as edicts, wills, and birth records. Although all tablets are rectangular, they differ in their proportions. To judge from surviving examples, school tablets could have as many as eight different "leaves," account tablets as many as seven, and legal tablets could have either two (typical in Egypt) or three. School tablets would normally have two hinging holes; legal tablets would have three; and account tablets, four, arranged in two distinct pairs. The number and arrangement of the sewing holes is important for our discussion because paired holes such as those in account tablets are also used later in papyrus codices. In all these tablets, the hinging holes were made along a long edge. The writing would run lengthwise in school tablets and across the width of account tablets, while legal tablets could combine both orientations: the main text, which was supposed to be sealed and thus not visible, would be written lengthwise, but a brief summary was written on the outer face across

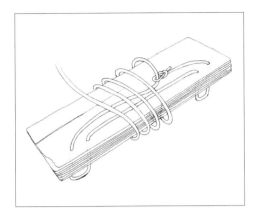

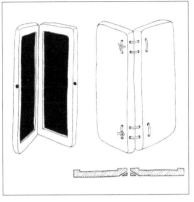

Fig. 9 (Left) Reco[...] of method for joi[...] wooden tablets of a farm account codex found at Kellis in the Dakhleh Oasis, ca. AD 360. See Cat. 5.

Fig. 10 (Right) Reconstruction of method for connecting two wax tablets. Redrawn and adapted from Wilhelm Schubart, *Das Buch bei den Griechen und Römern*, 16, fig. 1. See Cat. 4.

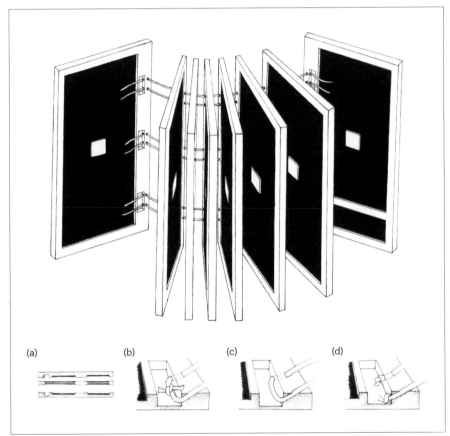

Fig. 11 Reconstruction of method for connecting eight wax tablets of an octoptychon from the House of the Bicentenary in Herculaneum: (a) cross section of the codex (showing only one inner tablet); (b–d) three possible methods for securing the thread connecting the tablets inside the cut recesses. See Cat. 3.

(a)   (b)   (c)   (d)

the width of the tablet. Nevertheless, there are exceptions, especially in the number and distribution of the sewing holes from the fourth century on. For example, the Isocrates wooden codex from the Dakhleh Oasis in Egypt has two pairs of sewing holes, much like the farm account codex dated from between AD 325 and 360 unearthed in the same place.[15] Beyond the texts contained, these differences in the number and arrangement of holes may have distinguished one type of tablet codex from another, and it would be interesting to know whether there were practical reasons for the differences.

Although Szirmai has noted that tablets could be connected "using hinges, metal rings or lacing," no scholar has considered the different methods in any detail.[16] We lack physical evidence that metal rings were used for this purpose during the Roman period, although such a simple hinge mechanism would have been efficient for codices with only a couple of tablets. For connecting the two tablets of a diptych, a thin cord would provide a simple alternative (see fig. 10), and in fact there are various methods by which two or more tablets can be connected together in a stable yet flexible way that allows for free opening and closing. One of these—and to my knowledge the only one to have survived intact on a wooden codex—is the method used for connecting the tablets of the two wooden codices from the Dakhleh Oasis (see fig. 9). With this simple yet effective method, individual tablets could be easily loosened, opened, and retightened for either writing or reading. This particular method, however, does not allow the tablets to open without loosening the slipknot, which loosens the whole structure.

A more sophisticated method for attaching several tablets can be found in two wax tablet codices from Herculaneum, one consisting of five tablets and the other consisting of eight.[17] The first was sewn through eight holes distributed along the spine in four pairs; the second, through six holes in three pairs (fig. 11). Several wall paintings from Pompeii show similar codices, in which the tablets appear to be connected with the same method (see, for instance, fig. 1).[18] In the pair of Herculaneum codices as well as in a few other similar examples from Egypt and from other Roman provinces, each pair of holes is sewn independently, securing the sewing strings inside the cut recesses, perhaps with one of the three methods shown in figure 11b–d.[19]

More options seem possible, particularly for those wooden codices that have slight recesses along the spine, aligned with the sewing holes, to accommodate and at the same time secure the thread connecting the tablets (fig. 12c).[20] Similar but much wider cuts were also used in the Vindolanda "leaf" tablets from the first and second centuries AD (fig. 12a) and also in legal tablets for recessing a cord to keep the tablets safely closed (fig. 12b).[21] This possibility, as well as the evidence of worn wood between the sewing holes that is clearly visible in the upper and lower covers of the Papnoution school tablet codex now in the Louvre (fig. 13), suggests that there was more than one way to connect a number of wooden tablets by passing cord through the holes and the spine recesses.

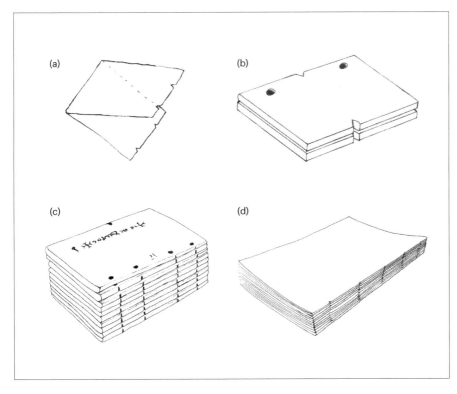

(a)

(b)

(c)

(d)

Fig. 12 Recesses cut in spines for sewing or securing thread in different writing devices. Clockwise: (a) letter from Vindolanda written on tree bark; (b) Roman legal tablet, closed; (c) Theodoros wax tablet codex from Fayoum, Egypt, 7th century AD, now at the Musée du Louvre in Paris, MNE 914; (d) parchment codex shown from spine with gatherings prepared for sewing.

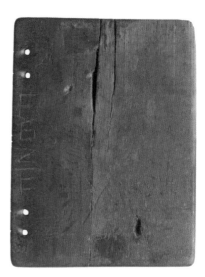

Fig. 13 Upper and lower boards of the tablet codex belonging to Papnoution, from Antinoe, Egypt, ca. AD 307–425. Wax and wood. Musée du Louvre, MND 552.

Four possible methods are shown in figure 14, but in practice some methods would have worked better than others. Figure 15 shows one of the easiest and probably most effective methods for doing so.

Regardless of the particular method, the principle of connecting several wooden tablets through holes is not unlike the principle of connecting individual sticks or canes. Methods such as binding, looping, and knotting, which have been known since ancient Egypt and used for making mats, baskets, and fences, could have been a source for the technique or techniques used in the wooden tablet codex.[22] The binding technique is well represented in the excavation of Qasr Ibrim, a fortified town in Lower Nubia in Egypt. The excavation strata date mostly from the third to sixth centuries AD, but the technique is still used in contemporary Nubia (fig. 16).[23] The method shown in figure 15 is almost identical to that in figure 16 except that in the wooden codex the thread (or cord) connecting the individual consecutive tablets is worked only around the tablets, while in the mats the sticks are sewn at a right angle on top of a supporting stick. As a rule, the four holes of the tablets are arranged in two distinct groups, usually far from each other, one toward

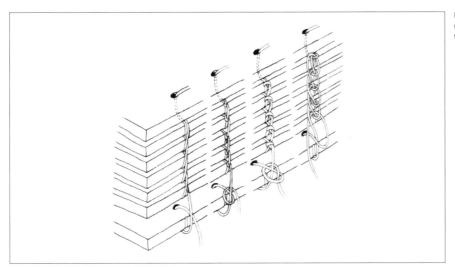

Fig. 14 Four possible methods for sewing wooden tablets together.

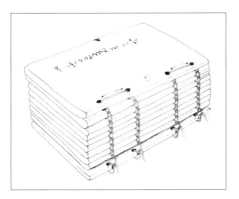

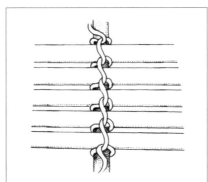

Fig. 15 (Left) Reconstruction of possible method for sewing tablets together, in this case the tablets of the Theodoros codex (see fig. 12c). The method is essentially a variation of the "binding" technique in basketry.

Fig. 16 (Right) Common basketry technique, called "binding," from late antiquity in Egypt for fastening several sticks with thread on an underlying stick. Redrawn and adapted from Wendrich, "Basketry," 256, fig. 10.1.i.

the head and the other toward the tail, which would imply that the tablets were sewn in two separate yet parallel sequences, with each pair of holes sewn independently (a feature already seen in the tablets from Herculaneum).

A similar process has been documented among some of the earliest bound single-gathering and multigathering codices to have survived and is used in the Ethiopic binding technique, which survived unchanged well into the twentieth century and is possibly still in use today. Of course, the possibility that the sewing was completed in a single sequence from one hole to the next through all four holes should not be excluded, but methods for doing so are certainly more complicated, not necessarily more effective, and have the disadvantage of exposing several lengths of the sewing thread along the spine between the sewing holes, creating potential structural weaknesses because their exposure makes them prone to breaking. In addition to creating a neater structure, the advantage of using sewing holes in independent pairs is that the binding will hold even if the thread breaks in one of them; only one pair of sewing holes would need to be repaired, not the entire structure.

Because our knowledge and understanding is conditioned by the random survival of physical evidence, some aspects of wooden tablet codex bindings remain uncertain. For instance, wall paintings from Pompeii, Herculaneum, and Rome, as well as many funerary stelae from Asia Minor, depict wooden tablets with a straplike appendage extending from the head edge of the spine.[24] The feature can be more or less clearly seen in three Roman wall paintings, the famous "Sappho" bust portrait from Pompeii (fig. 17), a

Fig. 17 Portrait of a young woman, "The Poetess of Pompeii" (Sappho?), from Pompei Insula Occidentalis, AD 50–70. Roman fresco. Museo Archeologico Nazionale, Naples, MN. 9084.

fragment of a wall painting from Herculaneum, and another fragment of a wall painting from the theater at Nemi, now preserved in the National Roman Museum (fig. 18). Although the subject requires further research, this appears to be some sort of device used for storing and carrying metal styluses along with the wooden tablet codex.[25] Although not related in any straightforward way, these extending straps invite comparison to the extension of the leather spine cover of the Glazier codex (Morgan Library and Museum, MS G.67, discussed in chapter 5 and shown in figures 45 and 46). In other examples, such as the two wax tablet codices lying open in front of Trebius Iustus in figure 2 (introduction), loops rather than straps are clearly shown extending from both the head and tail edges. Could these be some sort of device for hanging the tablets, as shown in a stela from Salona?[26]

Notes

Epigraph: Translation from http://www
.stoics.com/seneca_essays_book_2.html
#%E2%80%98VITAE1 ("On the Shortness
of Life," xiii.2–6).

1. Cribiore, *Writing, Teachers, and Students in Graeco-Roman Egypt*, 69. Cribiore considers 162 Roman and early Byzantine school tablets.
2. Meyer, "Roman Tabulae, Egyptian Christians, and the Adoption of the Codex," 304. Also see Cauderlier, "Les tablettes greques d'Égypte."
3. The main objection to the thesis that the multigathering codex evolved from the wax tablet codices can be found in Szirmai's "Wooden Writing Tablets and the Birth of the Codex," 31–32, and his *Archaeology of Medieval Bookbinding*, 3–4.
4. On the Uluburun diptych, see Payton, "The Uluburun Writing-Board Set"; Symington, "Late Bronze Age Writing-Boards and Their Uses." Written evidence suggests that wooden tablet codices in those early days could have as many as five tablets (Symington, 113). For chests and boxes with an elaborate system for connecting the boards and allowing them to open and close like that used in the Uluburun diptych (fig. 6), see McWhirr, *Roman Crafts and Industries*, 49, figs. 15 and 16.
5. For the use of tablets by the Hittites and the Greeks, see Roberts and Skeat, *Birth of the Codex*, 11–13; Regemorter, "Le codex relié à l'époque néo-hittite," 177; Howard, "Technical Description of the Ivory Writing-Boards from Nimrud" (describing an elaborate hinged system for connecting tablets that is very similar to that of the Uluburun diptych). On wooden tablets in general, see Lalou, *Les tablettes à écrire*; Blanchard, *Les débuts du codex*. Homer refers to wooden tablets in the Iliad (Z, 169). See also Atsalos, *La terminology du livre-manuscrit à l'époque byzantine*, 52. For Greek and Roman literary evidence on wooden tablets, see Degni, *Usi delle tavolette lignee e cerate del mondo greco e romano*, 73–146.
6. For wall paintings, see Meyer, "Writing Paraphernalia, Tablets, and Muses in Campanian Wall Painting"; for sculpture, see Türr, *Eine Musengruppe hadrianischer Zeit*.
7. For nonwax wooden tablet codices, see Sharpe, "Wooden Books and the History of the Codex."
8. Clarysse and Vandorpe, "Information Technologies," 719. For tablets with red wax, see Petrie, *Objects of Daily Use*, 66, cat. nos. 67–69.
9. In *On the Training of an Orator*, Quintilian (ca. AD 90) gives the following advice: "It is best to write on wax tablets, which provide the easiest method for erasure, unless relatively weak eyesight requires the use of parchment instead. Although parchment does aid visual acuity, it also delays the hand and breaks the train of thought because of the need to repeatedly remove the reed pen to refill it with ink." Quoted in Humphrey, Oleson, and Sherwood, *Greek and Roman Technology*, 528. Roberts and Skeat also quote the passage in *Birth of the Codex*, 21.
10. For examples of large tablets, see Millard, *Reading and Writing in the Time of Jesus*, 173, fig. 36, which depicts tablets being carried for burning on one of the two Anaglypha (or Plutei) Traiani from the Curia Iulia in the Roman Forum.
11. For notched edges, see Cribiore, *Writing, Teachers, and Students*, cat. no. 404, figs. 77 and 78 (fourth or fifth century AD). For tabs, see Capasso, "Le tavolette della Villa dei Papiri ad Ercolano," 223, fig. 2 and note 11.
12. An example of a five-tablet codex appears on Attic red figure kylix MS 4842, ca. 480 BC, in the University of Pennsylvania Museum. For an Egyptian seven-tablet codex, third century BC, see Petrie, *Objects of Daily Use*, 66, cat. nos. 67–73 and plate 59. Capasso refers to the possibility that wax tablet codices with as many as twenty-four tablets were found in a mid-eighteenth-century excavation at Herculaneum that have subsequently been lost ("Le tavolette," 222–225). For iconographical evidence on wax tablet codices with ten or more tablets, see Cavallo, "Le tavolette come supporto della scrittura," 97–104, figs. 1–4.
13. Martial epigrams 3, 5, 6, and 4, respectively. See Leary, *Martial Book XIV, The Apophoreta*, 58–62.

14. An interesting letter of unknown provenance written in Greek from the fourth century AD reads:

    > List addressed to my highly esteemed brother Daniel, perfume seller, by me Phoibammon, public scribe—in order, with the help of God, his Excellence, after his stay in the capital city of Alexandria, buy for me the following objects:
    >
    > An Antiochean robe, embroidered, little used, of a value of 10 keratia more or less;
    >
    > A small chair, made in the workshop, ink, an Antiochean kalamon, double, of a value of one and a half keration;
    >
    > A big square deltarion (i.e., codex) of ten tablets, with the tablets thin as leaves and in their middle a small wood so that the wax does not [text missing].

    Quoted in Scherer, "Liste d'objets à acheter à Alexandrie," 74. My translation.
15. Sharp, "Wooden Books and the History of the Codex." For more examples of school tablets with two pairs of hinging holes, see Cribiore, *Writing, Teachers, and Students*, cat. nos. 326, 327, and 329 (second century AD); cat. no. 389 (third or fourth century); and cat. no. 404 (fourth or fifth century).
16. Szirmai, *Archaeology of Medieval Bookbinding*, 3. Cribiore, *Writing, Teachers, and Students*, 65, note 74 refers to an eight-tablet codex (cat. no. 385) fastened together with a silk cord (the fastening method is not specified). For examples of tablets from between AD 325 and 375 that preserve part of their attachment, see Hope and Worp, "Miniature Codices from Kellis," 234. A small vestige of a broken leather strap used to connect the wax tablets of Chester Beatty codex WMS 142 (fourth century) is still preserved in one hole of the sixth tablet (WMS 142.6).
17. For the Herculaneum tablets, including a very clear drawing, see Pugliese Carratelli, "L'instrumentum scriptorium nei monumenti pompeiani ed ercolanesi," 270–278 and fig. 26. The tablets of the pentaptychon (five tablets) measure 4¾ × 1⅜ in. (12.1 × 3.6 cm);

those of the octophychon (eight tablets), 5⅜ × 2¾ in. (13.5 × 7 cm). In both, the thickness of the individual boards is ¼ in. (6 mm) for the outermost boards, which could also be considered as covers, and ⅛ in. (4 mm) for those in between.
18. For examples of codices depicted in wall paintings from Pompeii and Herculaneum, see Meyer, "Writing Paraphernalia," figs. 2, 3, 6, and 7; see also Mattusch, *Pompeii and the Roman Villa*, figs. 96 and 97.
19. Other examples with similar recesses for sewing the boards together are, for example, tablets 76 and 77 in Petrie, *Objects of Daily Use*, 66, 67, fig. 59. Both are Roman-period tablets from Egypt. Another example appears in Marichal, "Les tablettes à écrire dans le monde romain," 181, fig. 4, without any information about the date or the find spot. There are also three wooden tablets belonging to the same Roman-period codex in the British Museum (EA 26801). For examples from Vindonissa, a Roman legion camp in today's Switzerland, see Speidel, *Die römischen Schreibtafeln von Vindonissa*, 24, 29, 90–91, and fig. 11.
20. Tablet codices with such recesses can be found in various collections; because the recesses are usually not very deep, it is not always easy to identify them through photographs. Examples include a fourth-century and a sixth-century wax tablet codex (MND 522 H, known as the Papnoution codex, and MNE 914, known as the Theodoros codex) in the Louvre; a sixth-century tablet in the Morgan Library and Museum (M.1032); a sixth-century tablet in the Benaki Museum (cat. no. 618); a fourth-century tablet in the Chester Beatty Library (WMS 142.2); and a fourth- or fifth-century tablet in Berlin (T. Berol. 14000). To my knowledge, Cribiore is the only scholar who has noticed these cuts, interpreting them as "probably [made] to keep the strings in place and to prevent them from damaging the wood" (*Writing, Teachers, and Students*, 66).
21. Speidel, *Die römischen Schreibtafeln von Vindonissa*, fig. 5.

22. Although from a different part of the world, there are examples of Chinese bamboo books from as early as the fifth century BC that consist of strips of bamboo, sewn together through twining, with writing on their flat surface.

23. See Wendrich, *The World according to Basketry*, 110–111, 246–247.

24. On the two wall paintings, one from Pompeii and the other from Herculaneum, see Meyer, "Writing Paraphernalia," figs. 1 and 3. See also introduction, note 3. On the funerary stelae, see Waelkens, *Die kleinasiatischen Türsteine*, cat. nos. 31, 335, 346, 347, 354, 355, 405, 406, 439, 434 (second–third century). A first-century BC or first-century AD marble grave stela from Megara in Attica, now at the Benaki Museum in Athens, shows two papyrus rolls and a wooden tablet codex displayed in profile with the same appendage extending from the spine (erroneously identified as a box in the museum's explanatory label). The small star-shaped square seen on one of the boards at the middle of its fore edge implies a closing device and can also be seen on two wooden codices from Herculaneum and indeed on the external face of their back boards. See Pugliese Carratelli, "L'instrumentum scriptorium," 271, 278; Speidel, *Die römischen Schreibtafeln von Vindonissa*, fig. 12.

25. Elizabeth Meyer notes the strap extending from the spine of the wax tablet codex in the "Sappho" portrait ("Writing Paraphernalia," 586–587) and gives other examples. She identifies the stylus inserted in this device, but as she says, "no one truly knows what all this is."

26. For the stela from Salona, see Speidel, *Die römischen Schreibtafeln von Vindonissa*, 19, fig. 4.

Two

The Single-Gathering Codex

This medicine, written in this way, was found by Claudianos in a parchment notebook after the person who used it died.

—Galen, *De compositione medicamentorum secundum locos libri*, 2nd century AD

There appears to be no preserved physical evidence of papyrus or parchment notebooks in Roman antiquity before the second century AD, despite indirect literary evidence from the writings of Catullus, Suetonius, Horace, Persius, Galen, and various Roman legal authors about their use.[1] Martial, for example, in his much-quoted "To the Reader," introducing his books of epigrams, refers to *brevibus membrane tabellis* (small parchment writing tablets). In addition, his epigrams written around AD 84–86 refer to *pugillares* of parchment, either blank (epigram 7) or containing the work of various authors, such as Homer (epigram 184), Virgil (epigram 186), Cicero (epigram 188), Titus Livius (epigram 190), and Ovid (epigram 192).[2] Such notebooks, made of either parchment or papyrus, were used for all kinds of informal purposes—drafts of literary works, accounts, and record keeping—probably earlier than the earliest literary evidence we have about them. The writing on both parchment and papyrus notebooks could be erased and the notebooks reused as palimpsests.[3] There is no way to know how these notebooks were made or whether they differed from single-gathering and multigathering codices in terms of structure rather than content. There is important and illuminating iconographical evidence, however, that has so far escaped identification in a bronze-and-silver statuette of an artisan from the middle of the first century BC, supposedly found off the North African coast (fig. 19). The male figure wears a short tunic, which identifies him as an artisan, secured with a belt around his waist and a small "booklet" placed between his body and the belt. With only one exception, this "booklet" has been consistently described as wax tablets or as a diptych, both in the museum's online catalogue and in the various publications in which it is considered.[4] Nevertheless, the naturalistic representation, with the notebook slightly bent by the combined pressure of the artisan's body and the constriction of the belt, leaves little if any doubt that this is not a set of rigid wooden tablets but rather something made out of flexible material, such as papyrus or maybe even parchment. An even more interesting

Fig. 19 Statuette of artisan, late Hellenistic Greek, ca. mid-1st century BC. Bronze, silver. The Metropolitan Museum of Art, Rogers Fund, 1972, 1972.11.1.

feature is what may be a reinforcement strip along the spine edge. This strip, which cannot be connected with anything we know of wax wooden tablets,[5] suggests some sort of reinforcement for keeping the leaves of the notebook together the same way that, as we will see, similar strips of leather were used in the Nag Hammadi single-gathering codices, in which case they are called "spine strips."[6]

Despite the almost complete absence of physical and iconographical evidence of such parchment or papyrus notebooks, we can reasonably suggest that they would have been simple structures, much simpler than the multi-gathering codex. Structurally speaking, the single-gathering codex consists of a single unit made of a number of papyrus or parchment folios placed one on top of the other, folded in the middle, and sewn through the fold. Depending on the number of folded folios in the gathering, the fore edge could be uneven because the innermost leaves would tend to protrude. When the fore edge is trimmed to create a straight edge, the innermost leaves are narrower than the outermost by as much as 2 in. (5 cm).[7] Besides needing trimming, single-gathering codices made of papyrus also had an inherent mechanical weakness, their tendency to break at the fold, not because of the structure but because of the weakness of papyrus.[8] All the Nag Hammadi codices had leather or papyrus stays inside the centerfold as reinforcements so that the thread would not cut through the papyrus leaves (figs. 20 and 21).[9]

The earliest surviving single-gathering codices date from the second or third century AD.[10] Turner lists fifty-eight from between the third and fifth centuries, all written on papyrus rather than parchment and consisting of as many as 140 leaves.[11] The so-called Nag Hammadi codices are the single most important find of this type of codex, comprising thirteen bound codices found in 1945 near the village of Nag Hammadi, about thirty miles north of Luxor in Egypt (fig. 22). Based on waste material used for constructing the boards, the codices have been dated to the second half of the fourth century AD. Eleven are single-gathering codices and two (codices 1 and 13) appear to be composed of two gatherings each.[12] Although other single-gathering codices survive from the third and fourth centuries, they provide almost no information regarding their structure.

Fig. 20 Construction of Nag Hammadi codex, combining elements from different codices. Broken lines indicate the leather spine strip.

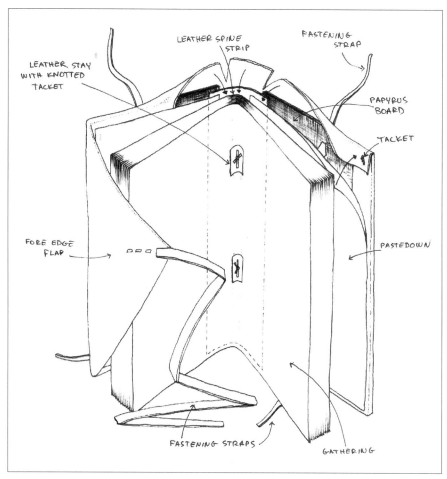

Fig. 21 Detail of tacket and stay used to sew a single gathering.

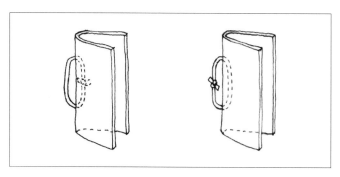

Fig. 22 The Nag Hammadi codices as they appeared when found in 1945. Coptic Museum, Cairo, Egypt. From Doresse and Mina, "Nouveaux textes gnostiques coptes découverts en Haute-Égypte."

Generally speaking, the making of these codices follows the same rather simple principle: two sets of holes are opened through the centerfold, one toward the head and the other toward the tail edge of the gathering.[13] Thread or string made of rolled parchment or papyrus passes through the holes and is knotted (or "tacketed"), either inside the centerfold or on the outside, with each pair of holes sewn independently. To prevent the papyrus leaves from tearing as they are sewn, leather stays (or guards) are added in the centerfold and the spine of the gathering is provided with a spine strip, which is also occasionally used for attaching the boards to the gathering (see figs. 20 and 21).

Tacketing itself has survived for centuries for keeping the leaves of the gatherings of multigathering codices together provisionally.[14] The spine strip of the artisan's notebook can be understood as functioning in the same way as the spine strips placed along the spine of the single-gathering codices. There is no detail to indicate sewing in the artisan's notebook spine, perhaps because of the small scale of both the statue and the technical detail itself. All the Nag Hammadi codices have papyrus boards made of manuscript waste and a leather cover. The boards can consist of either two separate pieces of papyrus laminate pasted at the front and back of the gathering through the extensions of the leather spine strip (see fig. 20) or a single piece of papyrus laminate, folded at the spinefold of the gathering and attached to it by sewing it with the gathering itself, which creates a stiffer, less flexible structure. A leather cover is pasted on top of the papyrus laminate, turned in on all sides (often V-shaped notches are cut at the spine to facilitate the opening of the gathering), and leather ties are laced through both the cover and the papyrus boards, usually on the head and tail edges of the two boards (figs. 22 and 23; see also fig. 20). Most of the codices have a fore-edge leather flap, either rectangular or triangular, with a long strap that was meant to be wrapped around the closed codex (fig. 22; see also fig. 20). Such long straps are clearly seen in several mosaics and paintings from late antiquity, as for example those in figure 103 (chap. 9).

There are also even simpler techniques to keep several leaves together, such as stab sewing (fig. 24) and overcasting (fig. 25). Stab sewing, one of the simplest ways to keep loose leaves together, usually requires opening

Fig. 23 Fresco from tomb of Cerula, Catacombs of San Gennaro, Naples, 5th–6th century AD. The open codices containing the four Gospels have fastening straps all around. Notice the two black stitches shown in the center-fold indicating the sewing. Catacombe di Napoli.

Part I: The Precursors of the Multigathering Codex

two pairs of holes, one toward the head and the other toward the tail, through the thickness of the book block (which could consist of single or folded leaves) at some distance from the spine, lacing a length of thread through them and around the spine, and then knotting the two ends of the thread together.[15] The technique is found in early codices, and its ease and effectiveness in keeping a number of leaves together led to its survival into much later periods.[16] Stab sewing is similar to the proposed method for sewing of multiple wooden tablet codices (see chap. 1 and figs. 14 and 15): the thread follows the same route through the thickness of the leaves—whether wood, papyrus, or parchment—and around the spine. Overcasting is another simple stitch, used since prehistory for sewing fabrics and leathers and reinforcing their edges. An early example of its use for sewing books is from the sixth or seventh century AD, a gathering of eight single leaves from the Monastery of Epiphanius (fig. 26). Used for sewing single leaves or repairing individual gatherings before sewing them with other leaves in multigathering codices, the technique has survived for centuries.[17]

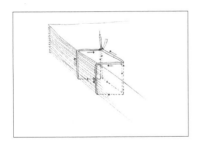

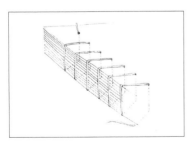

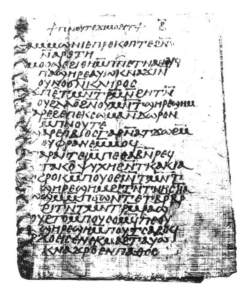

Fig. 24 (Top left) Stab sewing.

Fig. 25 (Bottom left) Overcasting.

Fig. 26 (Right) Simple codex of eight single leaves sewn with overcasting, from the monastery of Epiphanius, 6th or 7th century AD. From Crum, White, and Winlock, *The Monastery of Epiphanius at Thebes*, cat. no. 592.

Notes

Epigraph: I am grateful to Elias Tsolako-poulos for the translation from the original ancient Greek.

1. For an extensive discussion of the use of papyrus and parchment notebooks among the Romans, see Roberts and Skeat, *Birth of the Codex*, 15–23; see also ibid., plate 2 (bifolium from a leather notebook, second century AD).

2. Martial's "To the Reader; Showing Where the Author's Book May Be Published" (epigram 1.2):

   > You who are anxious that my books should be with you everywhere, and desire to have them as companions on a long journey, buy a copy of which the parchment leaves are compressed into a small compass. Bestow book-cases upon large volumes; one hand will hold me. But that you may not be ignorant where I am to be bought, and wander in uncertainty over the whole town, you shall, under my guidance, be sure of obtaining me. Seek Secundus, the freedman of the learned Lucensis, behind the Temple of Peace and the Forum of Pallas.

   From Martial, *Epigrams of Martial*, 24. For the other epigrams, see Leary, *Martial Book XIV: The Apophoreta*, 62–63, 249–253, 255–257.

3. On the term "palimpsest" and its use, see Roberts and Skeat, *Birth of the Codex*, 16–18.

4. The exception is Hemingway, "Statuette of an Artisan," 262–263. Although he identifies the booklet as a notebook rather than a tablet or diptych, he does not specify whether "notebook" indicates its format or its function. For a complete bibliography of the artisan statuette, see ibid., 262.

5. Boucher-Colozier notes the uniqueness of the strip in "Un bronze d'époque alexandrine," 26.

6. Szirmai, *Archaeology of Medieval Bookbinding*, 10–11, fig. 1.3.

7. Turner mentions three single-gathering codices in which the width of the leaves between the innermost and the outermost varies from a half inch to nearly two inches (1.5–5 cm). See *The Typology of the Early Codex*, 23.

8. For example, the single-quire codex of Menander (P. Bodmer IV) from the third or fourth century was repaired and resewn twice. Ibid., 57.

9. Parchment stays were also used in Chester Beatty Cpt. 2019.8, AD 300–350.

10. See Turner, *Typology of the Early Codex*, 59.

11. Ibid., 58–60.

12. Ibid., 60; Doresse, "Les reliures des manuscrits gnostiques coptes," 27–49. Unfortunately, it is unclear how the two codices with two gatherings were sewn, although according to Turner, codices I and XIII "are in fact 2 single-quire codices bound together" (*Typology of the Early Codex*, 60).

13. For a more detailed account of sewing single-gathering codices, see Szirmai, *Archaeology of Medieval Bookbinding*, 7–14.

14. For tacketing, see Petherbridge, "Sewing Structures and Materials," 376–378. Tacketing was also used extensively in medieval and later Western Europe; see Szirmai, *Archaeology of Medieval Bookbinding*, 287–290.

15. For stab sewing, see Petersen, "Coptic Bookbindings," 7–9.

16. Examples of stab sewing are codices Morgan M.202 and British Library Pap. 126 (both single-gathering codices containing Homer's Iliad, fourth century). See Szirmai, *Archaeology of Medieval Bookbinding*, 13; Petersen, "Coptic Bookbindings," 7–14. For a later Byzantine example, see Petherbridge, "Sewing Structures and Materials," fig. 6. The traditional binding technique in China and Japan is based on the same principle of stab sewing.

17. For the eight single leaves sewn with overcasting, see Winlock and Crum, *The Monastery of Epiphanius at Thebes*, pt. 2, 309–312, and plate 1. On the use of overcasting in Byzantine and post-Byzantine bookbindings, see Boudalis, "The Evolution of a Craft," 768–769.

Part II

The Multigathering Codex

Three

The Multigathering Codex: Introduction

There are twenty-two books which are too bulky to bind into one volume. If you want two volumes they must be divided so that one volume has ten books and the other twelve.

—St. Augustine, *Letter to Firmus*, AD 426–428

The multigathering codex, like a modern book, consists of leaves—papyrus, parchment, or paper—arranged in groups of usually four folios, which are folded in the middle to form a gathering of eight leaves and sewn along and through the fold with other gatherings. The sewn gatherings form the book block, which is usually provided with two boards, upper and lower, a spine lining, a pair of endbands sewn at the head and tail of the spine, then a cover, usually made of leather, and finally a number of fastenings. As a rule, a codex was first written and then bound because it is much easier to write on single leaves than on the leaves of a bound book; also, once a codex is bound, there is no easy way to add more than a couple of leaves, and it is not practical to remove any unused leaves. The multigathering codex is a further evolutionary step, combining features of wooden tablet and single-gathering codices. Part II will discuss the components and process of making multigathering codices in late antiquity, as well as the cultural, religious, and technological contexts within which they were made.

One of the most interesting and stimulating recent articles about wooden tablets and their role in the evolution of the codex is Elizabeth Meyer's "Roman Tabulae," which compares the dimensions of all three types of wooden tablet codex—school, account, and legal—along with other characteristics, such as the direction of the writing and the use of abbreviations and punctuation. Meyer concludes that the papyrus codex evolved from the legal tablet diptychs of Roman Egypt.[1] The argument is well researched and considered, but even if we accept that the features of the papyrus codex are borrowed from legal tablets, the number and arrangement of the sewing holes are more likely to have derived from account tablets, which normally have two pairs of hinging or sewing holes, as do most early single-gathering and multigathering codices. Account tablets, after all, were meant to be opened and closed regularly, unlike legal tablets, which were normally sealed to protect the authenticity of the documents contained within them. Compared to the types of codex seen so far, the multigathering codex is a more complex structure, combining several components and processes that were all, with the exception of endbands, also used in one form or another in the two earlier types of codex. Variations in the different components and processes distinguish (at least by convention, given that our knowledge is limited and often fragmented) one binding tradition from another in the wider context of the Mediterranean codex.

Concerning physical evidence, Szirmai considers in detail eleven more or less complete codices from between the fourth and seventh centuries, all of which contain Christian texts written on parchment and in Coptic, with the exception of the fourth to fifth century Greek Freer Gospel (Codex Washingtonensis, Freer Gallery of Art, F1906.297). All were found in Egypt in a Coptic context, although not in the same place. Because of their structural affinities, scholars view them as a more or less homogeneous group of early codex structures.[2] Another important group of early codices—containing classical Greek and Christian literature, written in both Greek and Coptic between the second and fifth centuries AD—is now shared mostly between the Bodmer Library and the Chester Beatty Library. Some of these codices still preserve substantial evidence of the techniques with which they were bound.[3] At least one, P. Bodmer XIX (fourth or fifth century), preserves enough of its structure to allow it to be included in the group of eleven codices that Szirmai lists. A few more isolated codices from this early formative period are also important, such as the Ambrosianus Syriacus C 313 inf (sixth or seventh century)[4] and the St. Cuthbert Gospel (early eighth century).[5] As with single-gathering codices, especially for the first centuries of the Christian era, there are hundreds of early multigathering codices surviving in a fragmentary state but with almost no substantial evidence of their binding structures.

The available physical evidence increases rather dramatically in the ninth and tenth centuries. Among the substantial number of bound Coptic codices preserved from this period, the two most well-known groups are represented by the fifty-two held by the Morgan Library and Museum, known as the Hamuli codices, originally part of the library of St. Michael Monastery in the Fayum valley, and those of the British Library, known as the Edfu collection, originally part of the library of St. Mercurios Monastery at Edfu in Upper Egypt.[6] Theodore Petersen, in his detailed study of the Coptic bindings in the Morgan Library, lists and describes another fifty Greco-Egyptian and Coptic bindings in other European, American, and Egyptian collections.[7] An important yet unspecified number of bound codices from the ninth and tenth centuries and possibly earlier periods preserved in the St. Catherine's Monastery library in Sinai still await proper identification, study, and publication. Unlike most of the early bound codices preserved in American and European collections that have been repaired during the twentieth century, the Sinai collection is in relatively pristine condition and is likely to cast light on various aspects of early bookbinding techniques.[8]

From the period between the ninth and tenth centuries, 77 Islamic manuscript bindings have survived out of a total of 179 bindings found in the Grand Mosque in Kairouan, Tunisia, while fewer bindings from the ninth and tenth centuries have also been found in Sana'a and Damascus.[9] Partly from normal wear and tear and partly from extensive past conservation treatments, the condition of most pre-tenth-century bindings often does not allow us to fully understand all their technical features.

In the chapters that follow, the major components of a multigathering codex will be considered in turn, following the actual sequence that a bookbinder would have followed to bind together a number of gatherings into a codex.[10] In addition, the specific techniques used in codex binding will be compared with those found in other crafts of late antiquity.

Notes

Epigraph: *Corpus scriptorum ecclesiaticorum latinorum*, Ep. 1*A; translation from Gamble, *Books and Readers*, 134.

1. Meyers, "Roman Tabulae."
2. Szirmai, *Archaeology of Medieval Bookbinding*, considers the eleven codices in a single chapter, "The First Multi-Quire Coptic Codices" (15–31). See also Sharpe, "The Earliest Bindings with Wooden Board Covers."
3. For the Bodmer papyri, see mostly codices P. II, V, VIII, XIII, XVI, XVII, XVIII, XIX, XXI, and XXIII. The Bodmer Library published its collection of codices between 1960 and 1965. On the Bodmer collection and various issues related to it, see the contributions in *I Papiri Bodmer*. For the Chester Beatty codices, see Kenyon, *The Chester Beatty Biblical Papyri*.
4. See Petersen, "Coptic Bookbindings," item 58, 305–309, and "Early Islamic Bookbindings," 53, note 11, fig. 21.
5. On the St. Cuthbert Gospel, see Pickwoad, "Binding."
6. On the Edfu collection of codices, see Lindsay, "The Edfu Collection of Coptic Codices."
7. On the Hamuli codices, see Depuydt, *Catalogue of Coptic Manuscripts in the Pierpont Morgan Library*; Petersen, "Coptic Bookbindings"; Needham, *Twelve Centuries of Bookbinding*, 12–13.
8. The St. Catherine's Monastery collection of codices has been surveyed in detail as part of the St. Catherine's Library Project, jointly undertaken by the St. Catherine's Foundation and Camberwell College of Arts. For further details, see the project website http://www.ligatus.org.uk/stcatherines/node/114.
9. On the Kairouan bindings, see Marçais and Poinssot, *Objets kairouanais*; Petersen, "Early Islamic Bookbindings." On bindings from Sana'a and Damascus, see Dreibholz, "Some Aspects of Early Islamic Bookbindings"; Deroche, "Quelques reliures médiévales de provenance Damascaine."
10. The issue of how gatherings were composed or constructed will not be treated here. On this issue, see, for instance, Wouters, "From Papyrus Roll to Papyrus Codex."

Four

The Sewing of the Gatherings

You did not respect me and you treated me like a dog and you didn't make for me the small metal needle which I am missing for the finishing (binding) of the book.

—Ostracon with a letter written by monk Frange, late seventh–early eighth century AD

As we have seen with single-gathering codices, there are several simple techniques for keeping multiple leaves together. For keeping the gatherings of a multigathering codex in a specific sequence, however, the sewing techniques inevitably need to be more elaborate. There are two broad categories: unsupported and supported sewing. In unsupported sewing, the gatherings are sewn together with a thread that passes through the centerfold of gatherings of the book block and then loops around the sewing of the previous gathering. In supported sewing, the thread that goes in and out of the gatherings also winds around cords or leather thongs—called sewing supports—arranged along the outer face of the spine folds, which are subsequently used to attach the sewn book block to the boards. Unsupported sewing, the earlier of the two methods, was used in all bookbinding traditions around the Mediterranean, in both the East and West, until the eighth or ninth century (fig. 27).[1] In traditional

Fig. 27 Sinai codex Greek 1155, with spine exposed, revealing underlying unsupported sewing structure of the book block, 16th century AD. St. Catherine's Monastery, Sinai, Egypt.

Byzantine and post-Byzantine bookbinding, it survived unchanged until the late seventeenth century and until the twentieth century in the Islamic tradition, while in traditional Ethiopic binding it was certainly used until the last century and possibly still survives today.[2] For the Syriac and Georgian traditions, the published data are too meager to allow for safe chronological conclusions, but broadly speaking, they seem to correspond to the chronology of the Byzantine and post-Byzantine tradition.[3] In all these traditions the abandonment of the traditional bookbinding technique was essentially triggered by the adoption of supported sewing. Supported sewing was apparently invented—initially in the form of "herringbone" sewing—around the eighth century in Western Europe for Carolingian bindings and has been used in the Western tradition ever since for books bound by hand (see fig. 37b).[4]

In medieval Western Europe unsupported sewing was used for limp bindings and from the late fifteenth century onward for printed books. From the mid-nineteenth century to the present, the technique is still consistently used in machine sewing the gatherings of printed books.[5] The earliest surviving codex with its original binding from Western Europe is the St. Cuthbert Gospel from Northumbria in North East England, which was sewn with unsupported sewing in four sewing stations arranged in pairs, with each pair sewn independently.

The actual process, which as a rule proceeds from one end of the book block to the other,[6] can be described as follows: Along the spine of each gathering, a number of sewing stations (at least two but usually three to five) are distributed in a regular or symmetrical way, each one marked by small V-shaped nicks, cut through the thickness of the spine edge of all gatherings.[7] The V-shaped nicks of the sewing stations of all the gatherings of a book block are aligned and therefore form recesses that run the width of the spine. These recesses can be compared to the very shallow cuts that we have seen in some wooden tablet codices and in some tablets from Vindolanda, in both cases used for the same purpose: to accommodate, recess, and protect the thread or cord that holds the tablets together (fig. 28; see also fig. 12, chap. 1). In several early codices the sewing stations are arranged in pairs—one toward the head and the other toward the tail—in a way similar to those in wooden tablet codices, school and account tablets in particular, and in single-gathering codices.[8] The same arrangement of sewing stations in pairs and the practice of sewing each pair independently had been preserved until recently in the Ethiopic binding technique and has also been used in machine sewing for binding books since the mid-nineteenth century.

The outermost sewing stations are called change-over stations because the sewing proceeds through them from one gathering to the next. Although the beginning and ending of the sewing may vary, the basic principle is the

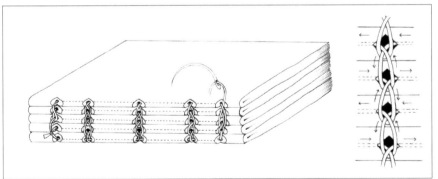

Fig. 28 Typical form and technique of unsupported sewing with loop stich. Broken lines indicate movement of thread in centerfold of gatherings. Inset to the right shows detail of the loops formed in the sewing process (direction of sewing from bottom to top).

same: the thread enters through either the head or the tail change-over station from the spine to the centerfold of a gathering, proceeds to the adjacent sewing station, exits through the V-shaped cut, drops, loops behind the loop of the previous gathering in the same sewing station, climbs, enters through the same V-shaped cut, and proceeds to the next sewing station.[9] As the thread exits from a sewing station and reenters after looping around the loop of the previous gatherings, it crosses over itself. The process continues until all the gatherings are sewn together, as shown in figure 28. In bookbinding literature as well as in everyday technical bookbinding terminology, this specific sewing technique has been designated as chain stitch or link stitch, but as will be seen in the discussion of socks made with cross-knit looping, the term "loop stitch" is more accurate.

Variations on this basic technique include the type of the stitch in the change-over stations, the number of sewing stations, the number of sewing thread lengths in the centerfold of the gatherings, the number of needles used for sewing, and the number of gatherings into which the sewing thread loops backward. Depending on the number of gatherings through which the thread loops backward, we can find one-, two-, or three-step (or more) loop stitch.[10] One of the major variations is the use of two lengths of thread in the centerfold of all—or only the outermost—gatherings of a book block, either using one needle that runs through the centerfold of the gatherings twice for each gathering or two needles working simultaneously (fig. 29).[11] In either

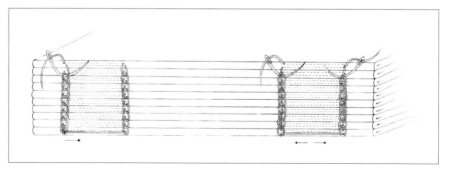

Fig. 29 Two methods for sewing gatherings of an early codex with independently sewn pairs of sewing stations. Left: sewn with single needle; right: sewn with two needles. In both cases, there are two lengths of thread—indicated with broken lines—between the sewing stations of each pair in the centerfold of the gatherings, a feature commonly found in several early codices.

case, the resulting sewing structure is significantly stronger than if it were sewn with only a single length of thread in the centerfold of each gathering. In any case, the thread always loops around the corresponding loop of the previous gathering and never passes through it. Because most of the earliest bound codices have been repaired and rebound, very little physical evidence of this type of sewing has been preserved. Thus, in the best cases we only have the descriptions of the persons, such as Lamacraft, who worked on their repair, and Petersen, who studied them at an early stage.

The visual characteristics of sewing across the spine and the technique itself invite comparison with similar techniques used in the context of contemporaneous fabric production. Chain stitch—as this ancient and widespread type of unsupported sewing is often called—does provide some visual similarity to the actual sewing technique used for the codices we are considering, but the similarity is only superficial and visual, not structural or functional. As a rule, chain stitch is used on already made fabric, and its purpose is decorative rather than structural—to embellish rather than create a fabric. It is formed by a linear sequence of loops in which one loop fits into the preceding loop; the direction of sewing is away from the pointed end of each loop (fig. 30; compare with fig. 28).[12] Furthermore, the thread in chain stitch does not cross over itself, a feature characteristic of the sewing of codices, as has been shown in figure 28. Chain stitch has been documented among Tutankhamun's burial fabrics, with late

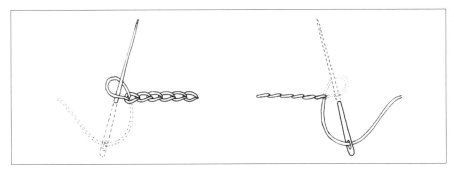

Fig. 30 Chain stitch.
Left: front; right: back.

antique examples from Palmyra in Syria (third century AD) and Egypt (fourth century and later). The stitch is much scarcer in Europe, at least until the high medieval period.[13] Chain stitch is common in Byzantine religious embroidered textiles, although none of the surviving examples can be dated earlier than the twelfth or thirteenth century.

Besides the chain stitch, which is sewn on an already made fabric, usually for decorative purposes, there is another technique, known as cross-knit looping, that is used to construct a fabric, according to Irene Emery, "by the repeated interworking of a single element [thread] with itself."[14] The technique has many names, including nålebinding, mesh stitch, ösenstitch, single needle knitting, encircled looping, knit-stem stitch, Coptic knitting, and looped needle netting.[15] Emery classifies this technique as a variation of simple looping, which is itself one of the earliest and most widespread fabric-making techniques;[16] however, cross-knit looping, in which the loop passes around the crossing of a loop in the previous row, differs from simple looping, in which the loop passes over the thread between the loops of the previous row (compare the three variations in fig. 31).

In its various forms, cross-knit looping is used in both fabric making and basketry. Emery points out that "the use of the [cross-knit looping] structure is so widespread chronologically and geographically that it is hardly an exaggeration to call it universal, and it ranges in application all the way from heavy rope fender covering and sturdy carrying nets to delicate, decorative laces. It is the basic stitch of needlepoint laces; it is used extensively for loose net-like structures and for firm, close-worked, cloth-like fabrics."[17] The technique has been used for making the stunning Paracas textiles in Peru (ca. 300–200 BC) and extensively in Africa.[18] It is apparently still used today in Iran[19] and also in Northern European countries and has been revived in recent years with the publication of several how-to videos on the Internet and in a number of books.[20] The earliest surviving examples of cross-knit looping can be dated as early as ca. 1000 BC.[21] Despite the wide distribution of the technique, the artifacts most closely related both chronologically and geographically to the sewing of the codex are socks from burials excavated in Egypt,[22] as well as a few fragments from the Dura-Europos excavation

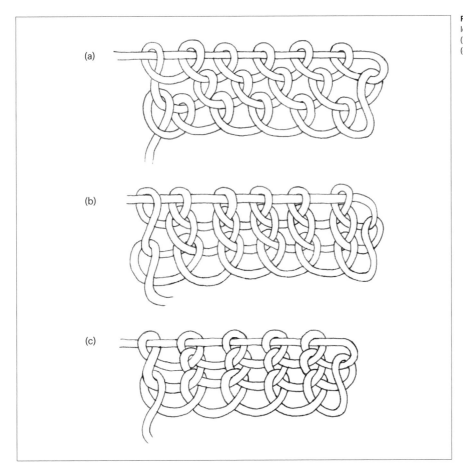

(a)

(b)

(c)

Fig. 31 Three variations of looping: (a) simple looping, (b) cross-knit looping, (c) pierced looping.

in Syria, the most elaborate of which (probably part of a dress), although originally classified as the earliest example of knitting, should probably be considered a more elaborate version of cross-knit looping (figs. 32–33).[23]

Dorothy K. Burnham's and Regina de Giovanni's experiments making socks with cross-knit looping show that they are worked with a limited length of thread and a blunt eye-needle, drawing or looping the full length of the thread through the adjacent loops (fig. 34). When the length of

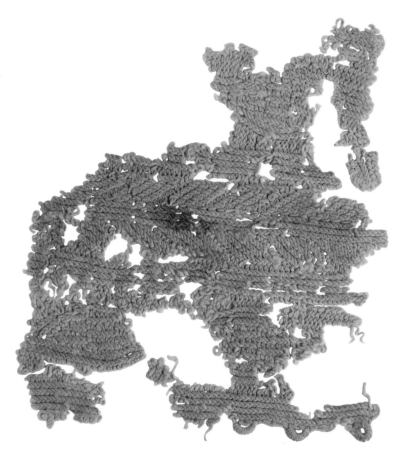

Fig. 32 Textile fragment made with cross-knit looping, Dura-Europos, ca. AD 200–256. Wool. Yale University Art Gallery, Yale-French Excavations at Dura-Europos, 1933.483. Cat. 12.

Fig. 33 Textile fragments made with cross-knit looping, possibly from a sock, Dura-Europos, ca. 323 BC–AD 256. Wool of three different colors. Yale University Art Gallery, Yale-French Excavations at Dura-Europos, 1935.556. Cat. 11.

Fig. 34 Modern replicas of an ancient sock and sock starter made with cross-knit looping. Acrylic, made by Regina de Giovanni. Cat. 13.

thread is consumed, a new length is added until the work has been completed. Such blunt eye-needles—usually made of bone—are common in museum collections of Roman artifacts.[24] The drawings accompanying Burnham's article show that the stitch can be worked equally well both spirally or back and forth from left to right and vice versa according to which part of the sock is being worked (fig. 35), as Regina de Giovanni also found in the process of replicating the socks shown in figure 34.

The visual and to some extent the structural characteristics of the technique for sewing codex gatherings with unsupported sewing are also similar to knitting (specifically, crossed knitting). In contrast to the limited thread used for cross-knit looping, knitting uses an "endless" or unlimited length of thread with which, according to Emery, "loops of a single continuous element [are] drawn through other loops already formed by the same element."[25] Thus, if one pulls the thread in a knitted fabric, it will unravel loop by loop, while in fabrics made with cross-knit looping (and also in a sewn codex), pulling the thread will have the opposite effect—that is, it will tighten the sewing.[26] The difference in the length of the thread between the two techniques is, according to Burnham, a clear indication that knitting is a much later development than cross-knit looping.[27]

The similarity between cross-knit looping and the sewing technique used in making codices is striking. In both cases, the actual movement of the

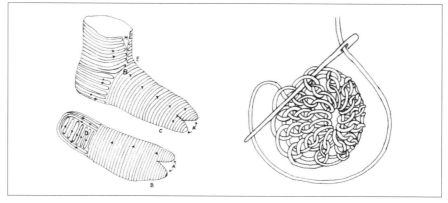

Fig. 35 Sock construction with cross-knit looping. Right: beginning of process; left: sock parts worked separately, then sewn together. From Burnham, "Coptic Knitting: An Ancient Technique," 123, figs. 3 and 4.

thread is exactly the same. Both use a short length of thread and a needle that in both cases works better if blunt. In both techniques, the pointed end of the loops indicates the direction of the sewing or looping (unlike knitting and chain stitch, in which the sewing proceeds in the opposite direction). In addition, both techniques can be worked back and forth. As in sewing gatherings, in which a one-, two-, or three-step (or more) loop stitch can be used, the number of steps in cross-knit looping can vary.[28] In chain stitch, the chains are formed one at a time in a vertical axis, while in cross-knit looping, several "chains" are built, which are aligned in both a horizontal and a vertical axis, a process similar to that of sewing a codex (see figs. 28, 30, and 31b). A further similarity is that in the sewing of the codex and in cross-knit looping, the thread in each loop crosses over itself, a feature completely absent in chain stitch proper (compare figs. 28 and 30). A feature of cross-knit looping can also help us better understand the mechanics of the codex. In socks made with cross-knit looping, the fabric has a tendency to curve on the outer face (the face on which the consecutive "chains" are formed). The tighter the cross-knit looping, the more likely the resulting fabric is to roll over.[29] Similarly, books bound with unsupported sewing have a well-known tendency to become concave when the sewing is very tight or the adhesive placed along the spine is very strong. Other stitches besides cross-knit looping produce looped fabrics. Although it seems that most of the late antique socks unearthed in Egypt were made with cross-knit looping, other examples show patterns formed with different looping stitches: a sock from the British Museum (EA72502) from between the fifth and seventh centuries AD, a sock in the Trier Museum dated generally to the Coptic period, and a sock from Saqqara from the sixth century, now at the MAK collection in Vienna.[30]

At this point, we could consider the sewing of codex gatherings as an adaptation of the cross-knit looping technique. Indeed, the structure of the codex may be understood as a fabric, in which the only difference from the original fabric-making technique is that in codex sewing the thread moves from one loop to the other by passing through the centerfold of the gatherings. Of course, the "chains" formed in the process are much more widely spaced than the closely packed "chains" formed by cross-knit looping in fabrics (fig. 36). As noted earlier, this sewing technique has remained virtually unchanged for centuries for all binding traditions of the Eastern Mediterranean, with the exception of Armenian bindings.[31] In Western Europe the history has been somewhat different, although loop-stitch sewing continued until at least the ninth century and even later for specific types of bindings.[32] The close connection between cross-knit looping—essentially an elaboration of looping and basketry techniques—and the sewing of codex gatherings invites another comparison. There is a similarity as well between

Fig. 36 Bands formed by sewing gatherings. Right: widely spaced bands of Sinai codex Greek 1155 (see fig. 27). Left: Photoshopped to show bands packed as in a fabric.

looping and basketry techniques around a foundation element that have been used from prehistory until the present in many parts of the world and the herringbone sewing of codices bound in Western Europe between the eighth and twelfth centuries. As shown in figure 37, a major difference between the two techniques is that in basketry the foundation element—the sticks around which the loops are made—runs through the loops horizontally, while in supported herringbone sewing, the support runs vertically. Whether the prototype for the sewing of the earliest codices lies in the making of socks or the prototype for the supported herringbone sewing of Carolingian codices lies in the mat and basketry technique of looping around a foundation element, the techniques used in the artifacts are structurally very similar. Using basketry terminology and classification, the sewing of gatherings on supports can in fact be described as a variation of a three-element binding.[33]

Although we cannot determine exactly how techniques for making socks and presumably other fabrics led to codex sewing, we can attempt to visualize it, placing codex making in a very mundane, perhaps domestic setting—a setting in which most of the earliest Christian literature was probably produced.[34] Petersen argues that "it might justifiably be assumed that it took a long time to elaborate and perfect that excellent technique, if it were not for the fact that at the time when the codex made its first appearance, numerous methods of lacing and interlacing threads and cord

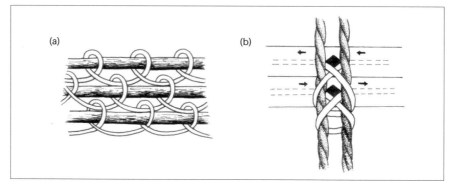

Fig. 37 Relation between (a) looping around a core, a mat and basket technique, and (b) herringbone sewing of Carolingian bindings. Broken lines indicate thread in centerfold of gatherings.

(a)

(b)

in fancy sewing, braiding, weaving, embroidering, and rug-making were already known in every textile shop of Alexandria and Rome. Thus the early bookbinders could readily recieve any advice needed on how best to lace things together."[35]

The earliest surviving examples of the multigathering codex structure can be dated as early as the second century AD,[36] but of these not enough survives to indicate how the leaves were fastened together. The earliest physical evidence of loop-stitch sewing in fully functional codices probably dates from the fourth or fifth century, at least two centuries after the earliest examples of multigathering codices.[37] In any case, loop-stitch sewing is inseparable from the innovation of the multigathering codex.

Sewing through the fold of the gatherings was not necessarily the only option for sewing together the gatherings of a multigathering codex. Stab sewing could also have been used either in paired sewing stations, each pair sewn independently (as we have seen in wooden tablet and single-gathering codices), using either one or two needles, as shown in figure 38, or in a

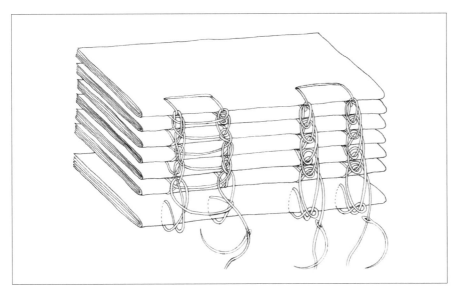

Fig. 38 Two hypothetical methods for sewing a multigathering codex with stab sewing, using one length of thread for each pair of sewing stations and either one needle (left) or two needles (right). Each pair of holes is sewn independently from the other.

single operation, sewing consecutively through all the sewing stations, as shown in figure 39. When we considered this last possibility for sewing the individual tablets of a wooden codex in chapter 1, we dismissed it as impractical because several lengths of thread would remain exposed in the spine between the sewing stations, but in the multigathering codex leather would cover the threads. Thus, sewing all the sewing stations in one session, as shown in figure 39, may have been an option. If we accept that stab sewing could have been used, in any of the variations just described, for multigathering codices, as the available evidence suggests,[38] then opening such a codex between two gatherings would reveal lengths of thread extending from the sewing holes to the spine parallel to the head and tail edges of the codex, as shown in figure 40.

Fig. 39 Hypothetical method for stab sewing all gatherings of a multigathering codex with one needle and one thread moving through all the sewing stations in a single operation.

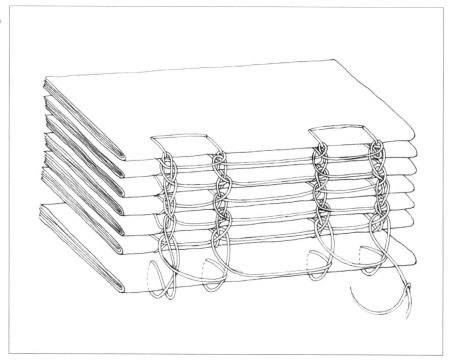

Part II: The Multigathering Codex

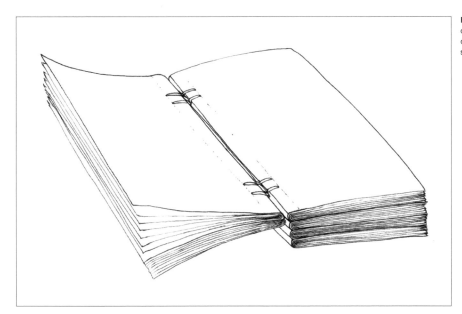

Fig. 40 Thread pattern in the opening between gatherings of multigathering codex sewn with stab sewing.

Surprisingly enough, this method can be compared with representations of open codices in mosaics, for example, the effigy of *Ecclesia ex gentibus* from the church of Santa Sabina in Rome from the fifth century AD (fig. 41, right) and three of the four open Gospel codices held by the four evangelists in the lateral walls of the presbytery of the Basilica of San Vitale in Ravenna from the first half of the sixth century (fig. 42 shows two of the four). In the Ravenna examples, the stitches are in pairs, as they are in the various surviving codices discussed earlier in this chapter. The open codex that Trebius Iustus holds in his lap similarly shows two pairs of sewing holes, although the sewing thread is not shown (see fig. 2, introduction). In *Ecclesia ex gentibus*, however, the codex is shown with only two stitches, set at a distance from each other, in the same way that the two open codicies shown in the fresco of Cerula are depicted (see fig. 23, chap. 2). The companion figure, *Ecclesia ex circumcision*, is also shown with an open codex, but it does not show any stitches at all (fig. 41, left), perhaps indicating a different contemporaneous book structure.

Fig. 41 Details of codices held by *Ecclesia ex circumcision* (left) and *Ecclesia ex gentibus* (right), 5th century AD. Church of Santa Sabina, Rome. The sewing stitches are visible on the book to the right.

Fig. 42 Details of the Gospel
codices of evangelists
Mark (left) and Luke (right),
first half of the 6th century
AD. Walls of presbytery
of Basilica of San Vitale,
Ravenna, Italy. Both open
codices clearly show sewing
stitches arranged in pairs,
one toward the head and the
other toward the tail.

Notes

Epigraph: Boud'hors, "Copie et circulation des livres," 158 and fig. 4. Ostracon, O. 292238. An ostracon is a potsherd, a cheaper material than papyrus, used for writing letters, receipts, and school texts.

1. On the use of unsupported sewing in both Eastern and Western Europe until the ninth century, see Szirmai, *Archaeology of Medieval Bookbinding*, 95–97.

2. On Byzantine and Ethiopic bookbindings, see ibid., chaps. 5 and 4, respectively; on the survival of unsupported sewing in the Byzantine tradition, see Boudalis, "The Transition from Byzantine to Post-Byzantine Bookbindings."

3. See Dergham and Vinourd, "Les reliures syriaques"; Kalligerou, "Tenth-Century Georgian Manuscripts."

4. On Carolingian bindings, see Szirmai, *Archaeology of Medieval Bookbinding*, 95–139.

5. On the use of unsupported sewing in limp bindings, see Szirmai, *Archaeology of Medieval Bookbinding*, 291–297; on the use of unsupported sewing in machine-sewn books, see Comparato, *Books for the Millions*, esp. 155–204; Gaskell, *A New Introduction to Bibliography*, 235–237.

6. The exception to this rule is double-sequence sewing, typical of Byzantine bookbindings, in which the book block is sewn in two sequences, each comprising one of the two boards and usually half of the gatherings. The two sewn halves are sewn together at the joint along the spine. See Pickwoad, *Assessment Manual*, 42. See also Petherbridge, "Sewing Structures and Materials," 398.

7. The term "sewing station" indicates the points along the spinefold where the sewing thread enters or exits a gathering in the sewing process. For a detailed account of the number of sewing stations in surviving early examples of codices, see Petersen, "Coptic Bookbindings," note 90.

8. Multigathering codices with the sewing stations arranged in pairs include Barcelona University Library P. Palau Ribes 181–183 (fifth century), St. Cuthbert Gospel (eight century), P. Bodmer XVI (fourth century),
and P. Bodmer XIX (fourth or fifth century). Other possible early examples from Western Europe are the Ragyndrudis codex, written in the eighth century and rebound around 1700, now in the Cathedral museum, Fulda, Germany; codex Würzburg University Library M.p.th.f.150, written in the eighth century; and Kassel GHB Th.f. 65, written in the sixth century (Szirmai, *Archaeology of Medieval Bookbinding*, 97, 291–294, fig. 10.6). As discussed in chapter 1, there are also school tablets with hinging holes arranged in two or more pairs.

9. The description of the process is based on the description of sewing through the folds in Spitzmueller's "A Trial Terminology."

10. On the variations of loop-stitch sewing, see Szirmai, *Archaeology of Medieval Bookbinding*, 16–18 and drawing 2.1. Szirmai calls these one-step, two-step and three-step link stitch.

11. On these sewing variations, see Szirmai, *Archaeology of Medieval Bookbinding*, 19–23, esp. drawings 2.3 and 2.4; Petersen, "Coptic Bookbindings," 25–36.

12. On chain stitch, see Emery, *The Primary Structure of Fabrics*, 241–243.

13. On the use of chain stitch in Ancient Egypt, see Vogelsang-Eastwood, "Textiles," 180. See also Schmidt-Colinet, Stauffer, and Al-As'ad, *Die Textilien aus Palmyra*, 30–31 and fig. 31; Jones, "A Historical Chain"; Mérat, "Étude technique et historique d'un ensemble," 131.

14. Emery, *Primary Structure of Fabrics*, 30.

15. For nålebinding, see Victoria and Albert Museum no. 2085&A-1900, http://collections.vam.ac.uk/item/O107787/pair-of-socks-unknown/. Some scholars, such as Hald, in *Ancient Danish Textiles from Bogs and Burials*, and Hansen in "Nålebinding," do not include cross-knit looping in the nålebinding technique. For mesh stitch, see Hald, *Ancient Danish Textiles*, 283–285. For ösenstitch, see Geijer, *Birka III*, 110, plates 31 and 32. For single needle knitting, see Victoria and Albert Museum no. 1936-1897, http://collections.vam.ac.uk/item/O128867/sock-unknown/, and no. 1243-1904, http://collections.vam.ac.uk/item/O128875/sock-unknown. For encircled looping,

see Seiler-Baldinger, *Textiles*, 17. For knit-stem stitch, see Bird and Bellinger, *Paracas Fabrics and Nazca Needlework*, 99–100. For Coptic knitting, see Burnham, "Coptic Knitting." For looped needle netting, see Hald, *Ancient Danish Textiles*, 283–285.

16. Emery, *Primary Structure of Fabrics*, 31. Seiler-Baldinger also classifies the technique under looping (*Textiles*, 10–18). For Stone Age examples, see Bender Jørgensen, "Stone-Age Textiles in Northern Europe," 1–10; see also Hansen, "Nålebinding," 21–27; Hald, *Ancient Danish Textiles*, 281–285 (examples from Denmark during the Neolithic period); Barber, *Prehistoric Textiles*, esp. 126–133.

17. Emery, *Primary Structure of Fabrics*, 31.

18. On the Paracas textiles, see Bird and Bellinger, *Paracas Fabrics*, and O'Neale, "Peruvian 'Needleknitting.'" On African textiles using cross-knit looping, see Warner Dendel, *African Fabric Crafts*.

19. Turnau, *History of Knitting before Mass Production*, 14.

20. See, for example, Claßen-Büttner, *Nalbinding*.

21. See Barber, *The Mummies of Ürümchi*, 32–33.

22. There are too many examples to mention them all. A sampling would include a pair of socks from the Victoria and Albert Museum dated AD 250–420 (http://collections.vam.ac.uk/item/O107787/pair-of-socks-unknown/); one sock dated AD 50–220 (http://collections.vam.ac.uk/item/O128867/sock-unknown/); one sock, fourth century, excavated in Hawara, Egypt (Tímár-Balázsy and Eastop, *Chemical Principles of Textile Conservation*); a sock from Oxyrhyncus, second century, now in Manchester Museum, University of Manchester (Schoeser, *World Textiles*, 62; a pair of socks at the Katoen Natie collection, carbon-dated, first–fourth century (Van Strydonck, De Moor, and Bénazeth, "14C Dating Compared to Art Historical Dating of Roman and Coptic Textiles from Egypt"); a child's sock, fourth–sixth century, in the Museo Egizio in Florence (Guidoti, *I tessuti del Museo Egizio di Firenze*, cat. no. 50); and one sock carbon-dated AD 200–400 at the British Museum (Pritchard, "A Survey of Textiles in the UK," 51 and fig. 17b). Cross-knit looping was not the only option for making socks; for a textile slipper/sock dated between AD 115 and 245, see Cardon, Cuvigny, and Nadal, "De pied en cap," 48–49. Outside Egypt similar fragmentary socks have also been found in Pompeii, Dura-Europos, and Masada, showing "that people were already 'knitting' and wearing socks in the first–third century AD and that their construction and design were relatively standardized" (Cardon, Cuvigny, and Nadal, "De pied en cap," 49). For a doll's hat from Egypt made with this technique, see Janssen, "Soft Toys from Egypt," 237–238 and fig. 8.

23. See Pfister and Bellinger, *The Excavations at Dura-Europos*, 54–56; Rutt, *A History of Hand Knitting*, 28–30, fig. 21. For a unique hairnet from Antinopolis, made with human hair using cross-knit looping, see Fluck and Froschauer, "Dress Accessories from Antinoupolis," 67.

24. A blunt needle prevents the needle from catching on the sewing thread. For examples of such needles, see McWhirr, *Roman Crafts and Industries*, 43, plate 20, showing bone needles found in London.

25. Emery, *Primary Structure of Fabrics*, 39. The terms "endless," "unlimited," and "indefinite" thread are used interchangeably because in knitting the loops are worked one through the other without the full length of thread passing through them as in cross-knit looping, and thus such thread can be "endless."

26. On knitting, see ibid., 39–42; Rutt, *A History of Hand Knitting*; Turnau, *History of Knitting before Mass Production*. Crochet is similar to knitting but developed later and is thus less relevant to this study. The provenance and history of knitting is uncertain, mostly because of the lack of early examples. The earliest surviving examples are stockings from Egypt from between 1200 and 1500 and late thirteenth-century cushions from the royal tombs in the Las Huelgas Abbey church in Spain. The earliest iconographical evidence in Western art, the so-called knitting Madonnas, are from the fourteenth century. Knitting may have been introduced in Europe through either trade routes from Asia or the Moorish invasion of Spain in AD 711–712. The complete absence of any artifact that could be iden-

tified as knitting needles provides indirect evidence that knitting was not employed in Roman and late antique fabric making.

27. See Burnham, "Coptic Knitting," 121.

28. On variations in cross-knit looping, see Emery, *Primary Structure of Fabrics*, 32; on variations in the sewing of codices, see Szirmai, *Archaeology of Medieval Bookbinding*, 17, fig. 2.1.

29. The rolled edges are clearly visible in the top edge of socks made with cross-knit looping, which has also been demonstrated by Regina De Giovanni, who kindly made the replicas of socks shown in fig. 34.

30. For the British Museum sock, see http://www.britishmuseum.org/research /collection_online/collection_object _details.aspx?objectId=118868&partId=1& searchText=EA+72502&page=1. For the sock from Triers, see Nauerth, *Die koptischen Textilien*, 105–106 and plate 13. For the MAK sock, which is somewhat similar in pattern and color to the British Museum example, see Noever, *Fragile Remnants*, 134, cat. no. 74. See also a pair of woolen socks from Antinoe vaguely dated in "epoca Copta," now in the Museo Archeologico Nazionale / Museo Egizio (AA.VV., *Antinoe*, 105, no. 107), and Noever, *Fragile Remnants*, 134, cat. no. 74, from sixth-century Saqqara. For a modern description of these different loop stitches, see Claßen-Büttner, *Nalbinding*, 21–25, 64–79.

31. Although the earliest surviving examples of Armenian bookbindings can be dated no earlier than the fourteenth century, they are characterized from the use of sewing supports, unlike any other type of binding from the Eastern Mediterranean at the time.

On these bindings, see Merian, "The Armenian Bookbinding Tradition in the Christian East"; Kouymjian, "Armenian Bindings from the Manuscript to Printed Book."

32. See Szirmai, *Archaeology of Medieval Bookbinding*, 101 and fig. 7.2, showing the spine of a ninth-century binding from the Abbey of Reichenau, with the gatherings and the boards sewn with loop-stitch unsupported sewing.

33. See Wendrich, "Basketry," 256, fig. 10.1, and 264.

34. In *Books and Readers*, Gamble writes that "early Christian manuscripts show that this literature was not commercially produced but was transcribed privately, that is, by Christians themselves. It can now be seen that this is no anomaly of early Christian texts: most ancient books were privately copied and circulated" (94).

35. See Petersen, "Coptic Bookbindings," 25.

36. See Turner, *Typology of the Early Codex*, 89–91.

37. One of the earliest examples of loop-stitch sewing in a multigathering codex, which is also preserved in astonishingly good condition, is the small (4¾ × 4⅛ in.; 12 × 10.6 cm) Glazier codex (Morgan MS G.67) from the fifth century. See Needham, *Twelve Centuries of Bookbindings*, 7–11; Kebabian, "The Binding of the Glazier Manuscript."

38. For specific examples of stab sewing, see Petersen, "Coptic Bookbindings," 10–14. Other examples include Morgan M.636, Ryland Coptic Pap. No. 7 and British Library Or. Ms. 3669 (both sixth or seventh century), and Chester Beatty Biblical Papyrus I (third century).

Five

The Boards and Their Attachment

You left me saying that you would bring me gum and old papyrus to the north. You have neglected to do so far. I trusted you, I cooked the glue: it has abated and went off; it is unusable. Now think to bring me the papyrus and the gum so that I can make the boards.

—Ostracon with a letter written by priest Moses, seventh- or eighth-century AD

Multigathering codices were normally provided with a set of boards made of some stiff, firm material to protect the book block and keep it compact. The boards of eastern Mediterranean codices were always flush with the book blocks until at least the late seventeenth century. Covers made with laminated manuscript papyrus waste were used exclusively for the Nag Hammadi codices and extensively for the boards of multigathering codices as early as the fourth century and possibly earlier. Up to the tenth century, most surviving Coptic bindings also have laminated boards, which can include not only papyrus leaves but various other components, such as folios of parchment manuscript waste, cloth, leather-tooled covers from discarded or rebound codices, papyrus and parchment off-cuts and trimmings, and straw mixed with adhesive. Such composite boards could be made to specific sizes, possibly using some kind of mold or cut to size from bigger boards.[1] Some of the boards are thick and heavy, such as the boards of the Morgan codex

M.599, made with a paste of straw mixed with an unidentified adhesive and possibly a filler between a half inch and an inch (1.5 and 2.4 cm) thick.

Besides composite boards, wood was the most extensively used material for covers. The earliest existing examples of wooden boards, preserved in the Chester Beatty Library, date from between the third and fourth centuries, although they cannot be related to any specific manuscript and pose various problems of interpretation and identification. The most elegant and finely crafted examples, Cpt. 801 (fig. 43), 802, and 803 (fig. 44), and another two, less elegant but certainly of the same construction (Cpt. 824 and 826), are identified as book boards because they cannot be convincingly identified as anything else and also because their proportions seem to correspond to the proportions of book blocks from the same period.[2] Although Cpt. 824 and 826 present holes that, based on later examples of bindings, can reasonably be considered fastening holes, in all of these boards the spine edge is shaped in a similar way, with a groove running the entire length from head to tail parallel with the spine. This groove might imply some method for attaching boards without using hinging thread. How these boards were connected to the sewn book blocks—if they were in fact book boards, as we suppose—is for the moment unclear, but they probably

Fig. 43 (Left) Wooden board with ivory inlays, possibly from a binding, 3rd or 4th century AD. © The Trustees of the Chester Beatty Library, Dublin, Cpt. 801.

Fig. 44 (Right) Wooden boards possibly from a binding, ca. 3rd century AD. Wood, ivory pins, and vestiges of leather and gilding. © The Trustees of the Chester Beatty Library, Dublin, Cpt. 803, 803a.

Part II: The Multigathering Codex

required the adhesion of some leather or textile extending from the spine of the book block.[3] There are also three pairs (Cpt. 809, 810, 811) and a single wooden board (Cpt. 812) from Egypt that have been vaguely identified as book boards, although I believe that their identification as such is doubtful.[4]

Some of the best-known examples of more or less complete codices date to the period between the fourth and seventh centuries. Except for Chester Beatty codex 815, which is bound in papyrus boards, the boards in other examples are wooden, and most present a peculiar structure in which the boards lack a cover, while the spine of the codex is covered with leather that extends on both sides and is pasted on the inner face of the boards; a number of thin leather strips are also laced through this cover and the boards.[5] The leather strips can vary in number from four (P. Bodmer XIX and Morgan M.910) and five (the Glazier codex, Morgan MS G.67) to as many as thirty-eight on Chester Beatty codex 814, a small codex with leaves measuring $4\frac{3}{4} \times 4$ in. (12.1 × 10.2 cm). Among all these bound codices, the only one that seems to be clearly understood in structural terms is the Glazier codex (figs. 45 and 46). All the other codices, for one reason or another, have features that have not been clearly understood and explained, despite their following the basic structural principles of the Glazier codex.[6]

There are many questions about this method of attaching the boards. Why would someone lace all these thin strips through the wood, spacing them so closely that the wood itself might split (as it has in one of the boards of Cpt. 814)? And why would someone rely only on adhesive to attach the spine cover and thus the boards to the book block, a method that might indeed work for small codices but would certainly have been completely inadequate for larger, heavier codices?[7] In addition, after using strips of leather to attach the leather spine cover to the boards, why was the further step not taken of using these strips as sewing supports, as was done later in Western European bindings to create a connection that was mechanically more effective? Why were sewing supports not invented earlier? After all, cross-knit looping is closely related to other looping techniques used for making baskets and mats, in which the actual thread winds around a core or foundation element, such as grass and reed (see fig. 37a, chap. 4). And the

Fig. 45 (Left) Modern facsimile of 5th-century AD Glazier codex (Morgan MS G.67). Made by Ursula Mitra. Cat. 19.

Fig. 46 (Right) Open view of figure 45. Unlike the example shown here, in the original codex the leather cover would have been adhered to the spine of the book block.

endbands found in later Coptic and related bindings extend and are sewn onto the boards to act as sewing supports. In other words, to judge from the surviving examples, the techniques used to bind books in the Eastern Mediterranean seem to have evolved more slowly than the corresponding techniques in Western Europe after the eighth and ninth centuries. In any event, the use of thin leather strips to attach book boards appears to have been completely abandoned, and no apparent evolutionary pattern connects it to the "thread hinging" methods, as Petersen called them,[8] that we find in codices from the eighth century onward. Should we view the leather strips as just one of several methods that would have been in use at the same time around the Mediterranean but was ultimately abandoned?

Methods for attaching the boards in later periods are well documented in original examples. The same techniques have been used for many centuries, until the seventeenth century at least, in Eastern Mediterranean bookbinding, with surviving examples from the Byzantine, Syriac, and Georgian traditions. In all these later codices, thread is used to connect the boards to the book blocks (thus the term "thread hinging") either in a separate process, before or after the book block has already been sewn, or in a process integral to the sewing that would start with sewing one board, then sewing the book block, and ending with sewing the opposite board in a single sequence.[9] Unless the threads used for sewing the boards and the book blocks are clearly different, as they are in figure 47, there is no easy way to distinguish between these two different techniques without visual access to all parts and elements of a bound codex, something that usually occurs only with badly preserved codices or codices undergoing conservation.[10] In the process of moving from one attachment station to the next, the thread can create different patterns between the attachment holes on the inner or outer face of the boards, usually a Z or U pattern and more rarely an X (fig. 48a–c). The same attachment

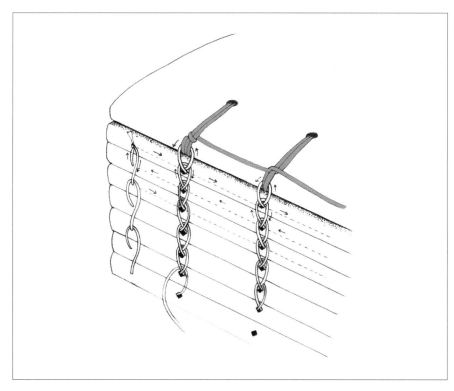

Fig. 47 One of several ways to sew a book block to the boards with thread hinging. In this case the boards have been previously prepared with a thread as shown in figure 49b, and subsequently the book block was sewn on the prepared board.

Fig. 48 Inner face of boards. Bindings of Morgan Library and Museum Coptic codices: (a) M.588, AD 9th century, (b) M.597, AD 10th century, and (c) M.634, AD 9th–10th century. From Petersen, "Coptic Bookbindings in the Pierpont Morgan Library," figs. 25–27.

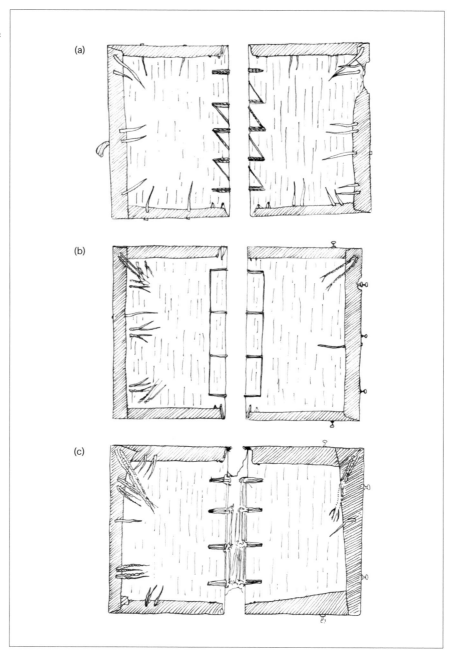

patterns can occur in both integral and separate board attachment methods and also with different thread routes.[11] As we have seen in the different techniques of cross-knit looping and knitting, patterns that look similar or even the same can in fact result from different techniques.

The thread lacing on the boards is structurally very similar to the blanket stitch—one of the simplest and oldest stitches used for textile edge finishes (fig. 49). The same zigzag thread pattern is seen in some rather uncommon papyrus documents, called double documents. They consist of a text written twice on the same sheet of papyrus, with one of the two copies of the text folded and "sealed" in the presence of witnesses to avoid alterations, while the other half remained visible. The earliest examples date from the fourth century BC, but the best-preserved examples are those found in the Cave of Letters in the Judean Desert from the first and second centuries AD (fig. 50).[12]

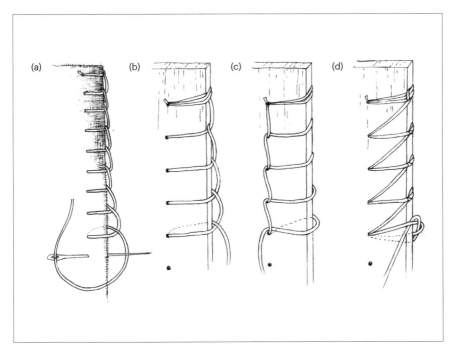

Fig. 49 Blanket stitch in (a) fabric; (b–d) variations of thread hinging boards of Eastern Mediterranean book structures. Compare with fig. 48.

(a)   (b)   (c)   (d)

Fig. 50 Papyrus double document, written December AD 127. Courtesy Israel Antiquities Authority.

Although there is no detailed description of the exact thread route provided in the publications concerning these documents, judging from the published photographs it is likely that the folded half of the document was secured with a variation of blanket stitch that creates the characteristic Z pattern shown in figures 49d and 50. Structurally speaking, these techniques, primarily employing thread to connect the boards to the book blocks—further enhanced by cloth spine linings extending onto the boards, sewn endbands, a leather covering, and attached pastedowns—are much more efficient and simple to make than those with multiple thin leather strips laced through the boards.

We know of other options for attaching book boards, and others certainly must have existed for which we have no evidence. Among the known options is a type of board that we could call integral because no separate technique is needed for attaching it to the book block: it is made of papyrus or parchment leaves that are folded and sewn at the two ends of the book block together with the text gatherings and subsequently covered with leather.[13] So-called double boards represent another common variation of board construction in multigathering codices. They consist of double thicknesses—an inner board connected to the book block through thread hinging and an outer, usually thicker board pasted on top of it. The extensions of the textile spine lining are pasted on the inner boards, the edges of the boards are covered with leather edging strips, while the outer boards are completely covered with leather, often richly decorated, before they are pasted on the inner boards of the book (fig. 51).[14]

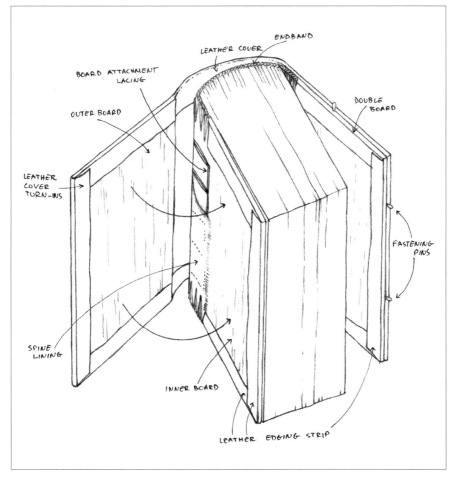

Fig. 51 Double-board binding structure. The boards consist of two thicknesses, the first attached to the book block with thread hinging, the second adhered to the first as part of the decorated leather cover. The fastening straps, which would have been laced through both layers of the left board, are not shown.

Notes

Epigraph: From Boud'hors, "Copie et circulation des livres," 158.

1. On the construction of the different boards in the bindings of the Coptic codices of the Morgan and other collections, see Petersen, "Coptic Bookbindings," 46–50 and note 117.
2. On the Chester Beatty boards, see ibid., 43; Regemorter, "Le codex relié depuis son origine jusqu'au haut moyen-âge" and "Some Early Bindings from Egypt at the Chester Beatty Library."
3. See Regemorter, "Some Early Bindings," 20.
4. On Chester Beatty boards Cpt. 809, 810, 811, and 812, see ibid., 12–16. Another board of the same kind is reproduced in Strzygowski, *Koptische Kunst*, 143, cat. no. 8820. See also *L'art copte en Égypte*, 66, cat. no. 40, from Antinoe, third–fourth century AD.
5. On this peculiar technique used for attaching the boards to a book block, see Szirmai, *Archaeology of Medieval Bookbinding*, 25. See also Sharpe, "The Earliest Bindings with Wooden Board Covers," 463, 468.
6. On the binding of the Glazier codex (Morgan MS G.67), see, for example, Needham, *Twelve Centuries of Bookbindings*, 7–11, esp. note 8.
7. The fourth-century Codex Sinaiticus, for example, originally consisted of about 730 parchment leaves, each measuring about 14 1/8 × 13 in. (36 × 33 cm). We must accept that other techniques would have been used for large codices, relying on mechanics rather than adhesive.
8. Petersen, "Coptic Bookbindings," 59.
9. The boards and the book-block gatherings of the eighth-century St. Cuthbert Gospel were also sewn in one sequence. See Pickwoad, "Binding," 48–49, fig. 2.8.
10. See, for example, Boudalis, "The Conservation of an Early 16th-Century Bound Greek Manuscript." Petersen, referring to the Morgan bindings, notes "that the sewing of a codex did not always include the hinging of the covers is clearly seen in several of the Morgan Coptic bindings, e.g., M.609, where the cord with which the manuscript was sewn is different in thickness and structure from the cord with which the covers were hinged" ("Coptic Bookbindings," 19, and note 78).
11. For an example of different attachment techniques that produce the same patterns, see Boudalis, "The Evolution of a Craft," 295–298 and fig. 93.
12. On these double documents, see Lewis, Yadin, and Greenfield, *Documents from the Bar Kokhba Period*, 1:6–10, and plates 12 and 24; Yadin, Cotton, and Gross, *Documents from the Bar Kokhba Period*, vol. 2, plates 34, 37, 38, 41, 42, 43, 44, 51, 52, 61, 62, 79, and 80.
13. On integral boards, see Szirmai, *Archaeology of Medieval Bookbinding*, 37, fig. 3.5b,d.
14. Bindings with double boards in the Morgan collection are M.569, 570, 574, 575, and 672. On these examples, see Petersen, *Coptic Bookbindings*, 59. In M.569 the inner half of the boards is 1/4 in. (7 mm) thick, while the outer is only about 1/8 in. (4.5 mm).

Six

The Spine Lining

The Lord will watch over the comings in and the
goings out of the house; for as long as the cross
is set in front of it the evil eye will not have power.

—Inscription on the lintel of a house at Sabba, Ethiopia, AD 547

B ecause the spine is as important to the anatomy of the sewn book block as it is to the human body, a great deal of attention has been given to protecting and reinforcing book spines in even the earliest examples of codices. Any material, usually cloth and more rarely parchment or leather, pasted to the outer face of the spine and used as a means for its protection is usually called a spine lining. The earliest examples suggest that this feature was originally used in single-gathering codices along with similar strips of leather or parchment in the centerfold to prevent the sewing or tacketing of the folded folios from tearing the papyrus (see figs. 20 and 21, chap. 2). In several of the Nag Hammadi codices, the strip of leather at the spinefold was used to attach the boards to the book block, indicating that the spine lining was also used to connect the boards or to reinforce their connection, a practice that persisted in Islamic bindings until early modern times (see fig. 20, chap. 2). Cloth spine linings are a standard feature of most Eastern Mediterranean bookbindings, with the exception of the Ethiopic tradition, in which case the spine of the book block was usually protected only with the leather cover. A spine lining normally consists of a single layer, but double or even triple layers appear to have been common, especially in Georgian and Syriac codices.

One of the earliest surviving examples of a multigathering codex with a cloth spine lining is Chester Beatty codex 815, a small sixth-century AD codex, measuring 3⅞ × 3⅜ in. (9.8 × 8.6 cm), bound in boards made of reused papyrus laminate, with a full leather cover. Although the boards have been completely eaten by insects, parts of the cover and spine lining are still preserved. Originally, the spine lining would have covered the full length of the spine, but because of its condition it is uncertain whether the lining extended onto the boards and, if so, how far. The sixth-century Chester Beatty codex 813 also originally had a spine lining made out of parchment, extending and pasted onto the inner face of both boards.[1]

The cloth used for the spine lining is usually a plain, rather coarse canvas-like textile, either in its natural beige color or often dyed blue (common among the bindings of the St. Catherine's Monastery library and the Morgan Coptic bindings).[2] In a few cases, it is possible to establish that the textile used was cut from some more elaborate fabric, possibly from liturgical or other vestments,[3] which may explain the embroidered or applied fabric crosses in the cloth spine linings of some codices. Such crosses appear to have been made as a sort of apotropaic or protective device (fig. 52). Examples can be found in a few Greek, Syriac, and Georgian codices from the St. Catherine's Monastery library and in a few other collections. It is likely that further examples of such talismanic spine linings have escaped notice because the spine lining is usually completely concealed under the cover of the codices.[4] Although the number of examples is small, their wide distribution in time and place would indicate a practice, so far unnoticed in bookbinding, that is well known and documented in fabrics (fig. 53; also see figs. 58 and 59), various objects, and buildings, in which the cross has been used extensively as a protective and a blessing symbol.[5] After all, as Haines-Eitzen says: "Books and bodies were vulnerable and the fact that pains were taken to protect both books and bodies alludes to their power."[6]

Notes

Epigraph: The inscription is quoted in Maguire, "Garments Pleasing to God," 218.

1. Lamacraft's description of Chester Beatty codex 813 is not very clear about the extension of the spine lining ("Early Book-Bindings," 220–221).

2. For example, blue spine linings from the Sinai collection are found in Sinai codices Greek 31 (Psalter, eighth–ninth century) and 213 (AD 967), as well as in Morgan MS M.574 (AD 897–898) from the Morgan collection.

3. An example with elaborate fabric is Sinai codex Georgian 18 (Lectionary, tenth century). The spine lining is a cutout, probably from a tunic; the piece of cloth includes a multicolored decorative woven band. Another example is Morgan MS M.609A, in which several colored threads are woven into the spine lining fabric.

4. The codices noted so far are in the St. Catherine's Monastery library: codices Greek 266 (Gospels, fifteenth century), Georgian 69 (Lectionary, thirteenth century), Syriac 44 (Menologion, eleventh century); in the National Library of Greece: codex 190 (Lectionary, early eleventh–early twelfth century); in the Iviron Monastery: codex 964 (Mathimatarion, AD 1562); in the Monastery of St. John the Theologian library, Patmos, Greece: codices 526 (various speeches, sixteenth century) and 904 (Triodion, fifteenth century). Also see detached binding British Library, Add MS 27860/1, probably seventeenth century.

5. For protective symbols in textiles, see Maguire, "Garments Pleasing to God," 218–219. Also Maguire, Maguire, and Duncan-Flowers, *Art and Holy Powers in the Early Christian House*, 18–22.

6. Haines-Eitzen, *The Gendered Palimpsest*, 9. On apotropaic devices on Armenian books, see Merian, "Protection against the Evil Eye?"

Seven

The Endbands

According to al-Isbīlī eight types of Byzantine (Greek) endbands (*al-ahbāk al-rūmīyah*) were known in his time. However, he describes only four of them, because the other four were more complicated and required demonstration.

—Adam Gacek, "Arabic Bookmaking and Terminology"

E ndbands are sewn bands at the head and tail spine edges of a book. Until as late as the seventeenth–eighteenth centuries AD, endbands in Eastern Mediterranean bindings were always sewn through the gatherings of the book block, usually around a cord core. Their purpose was to compress and "seal" the edges of the book block and the spaces between the gatherings, to reinforce the attachment of the boards, and eventually to decorate the book. Endbands probably appeared as the multigathering codex was introduced and gradually established (none have been found, nor are they necessary, for single-gathering codices). Out of the eleven codices dated between the fourth and seventh centuries AD that are preserved in more or less complete form, only the fifth-century Glazier codex (Morgan MS G.67) and Chester Beatty codices 813, 814, and 815 (sixth–seventh centuries) preserve—or preserved when originally found—some small vestiges of what are probably the original endband colored threads. Some indirect evidence can also be identified in the form of tie-down holes close to the head and tail edges of the spine in the Sheide codex (Princeton University Library, Sheide 144, fourth or fifth century), the Mudil codex (Cairo Coptic Museum, fifth or sixth century), and the Palau Ribes codex (Barcelona University Library, fifth century).[1] Thus, exactly how the endbands in these earliest codices were made is unknown, and there is still the possibility that these holes may have been made not for endbands but for the purpose of tacketing in order to keep the leaves of the individual gatherings together.

In chronological sequence, the next sound available evidence is found in the Hamuli codices preserved in the Morgan Library and Museum, generally dated in the ninth and tenth centuries, and in the tenth-century bindings of several Georgian codices preserved in the St. Catherine's Monastery library.[2] The endbands found on the Hamuli codices consist of small fragments that can nevertheless be identified as two different types: the first is the loop-stitch endband (Petersen's term is "Coptic endband");[3] the second is a twined endband. Besides the endbands in the tenth-century Georgian bindings, there are probably other equally early or even earlier endbands or vestiges of endbands in the St. Catherine's Monastery collection still awaiting proper identification and publication.

In the earliest bindings for which at least some evidence of endbands is preserved, the lack of any tie-down holes on the boards indicates that the endbands did not extend onto the boards but were used to protect only the head and tail edges of the book block. Paul Adam suggests a sewing technique in which the change-over stations were pushed into the head and tail edges rather than arranged along the spine toward the head and tail.[4] This would have created a loop stitch right at the edge rather than along the spine, as is normally the case (fig. 54; compare fig. 28, chap. 4), and thus would not allow the book-block edges to be trimmed to straighten them after they were sewn. The use of this technique is only speculative, but if it was ever used, it must have been abandoned early in the evolution of the codex and the change-over stations moved from the edges to along the spine, so that the edges of the book-block could be trimmed. Unfortunately, there is no physical evidence of the technique besides two

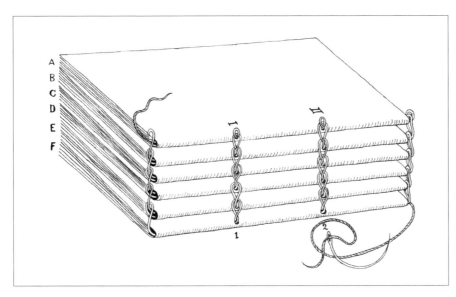

Fig. 54 Conjectural reconstruction of change-over sewing stations at head and tail edges, forming what could be described as the earliest endbands. From Petersen, "Coptic Bookbindings," fig. 17.

stitches on Chester Beatty codex 813 that, as Szirmai suggests, might only be misplaced sewing threads.[5]

From at least the ninth century onward, all endbands extended onto the boards and were fastened to them. With the exception of books bound with pasteboards in the typical later Islamic technique, this was one of the longest-lasting features of endbands in Eastern Mediterranean bookbinding traditions, and for good reason, because the extensions greatly reinforce the connection of the boards to the book block, functioning in structural terms as sewing supports do in Western bookbinding.[6] Although the use of endband extensions onto the boards started to decline in the late seventeenth and early eighteenth centuries, they can still occasionally be found in post-Byzantine bookbindings as late as the mid-nineteenth century.[7] The protrusion of the endbands beyond the edge of the boards, which created the characteristic rise at the head and tail edges of the spine, also declined in the same period and was completely abandoned by the early eighteenth century, when boards slightly bigger than the book blocks were adopted.[8]

As noted in chapter 4, the basic structure of the codex—that is, the sewing of the gatherings to form the book block—can be compared to a fabric structure into which the gatherings are incorporated through sewing. To someone familiar with fabric-making and embellishing techniques, it should be easy to make the connection between some of the endbands and certain textile edge-finishing techniques. Endbands are to books what edge finishes are to textiles: in both cases they are used primarily to compact and reinforce the edges of a fabric where it is vulnerable to wear and fraying, as well as for decoration.[9] The close connection between endbands and fabric edge finishes has been noted before. In 1913 Alfred Bel and Prosper Ricard described the endbands of thirteenth-century Moroccan bindings as similar to the edging in such garments as the burnoose and djellaba.[10] Textile edge finishes can be an integral part of fabric making or can be added after the fabric has been woven or cut to shape.

Although the literature on textiles is extensive, only a relatively small portion of the published research provides information relevant to the discussion of textile edging. Around the Mediterranean, one of the most common ways to work the edges of a textile was to create one or more bundles of warp threads at the edges around which the weft was woven.[11] The endbands described in the remainder of this chapter are among the earliest recorded, found in bindings from the St. Catherine's Monastery library and the Morgan Library and Museum. Although the types of endbands increase rather dramatically after the eleventh and twelfth centuries, this may only be a consequence of the greater survival rate of original bindings from these later periods.

Endbands are generally distinguished as simple or compound. Simple endbands are sewn directly on the head and tail spine edges of the book block, with one or more lengths of thread passing through the centerfold of gatherings. They can incorporate one or more cores but can also be sewn without a core. Compound endbands are made in two distinct operations, working the primary component and then the secondary component. The primary component is sewn directly on the spine edges and through the gatherings, while the secondary component is worked exclusively on the primary component without passing through the book block.

Loop-Stitch Endband

The loop-stitch endband is one of the most widespread simple endbands around the Mediterranean in the period up to the tenth century. Loop-stitch endbands have been documented for eleven Coptic bindings, most of them dated between the 9th and 10th century AD in the Morgan collection and among early Islamic bindings from Sana'a in Yemen, as well as in a number of Western European Carolingian and Romanesque bindings from between the eighth and fourteenth centuries.[12] The endbands are formed with loop stitch, sewn through the boards and the centerfold of each gathering so that a row of linked loops is formed at the head and tail edges (fig. 55). A similar kind of sewing appears to have been used for the endbands of the St. Cuthbert Gospel, which was bound in the early eighth century in England.[13] Loop stitch was used to reinforce the neck or other openings of late antique tunics from Egypt. Good examples can be found in the opening of the side vent of an Egyptian child's tunic from the seventh to eighth century AD, now in the Whitworth Art Gallery of the University of Manchester, and in an Egyptian tunic from the ninth to eleventh century in the Egyptian Museum of Berlin (fig. 56).[14]

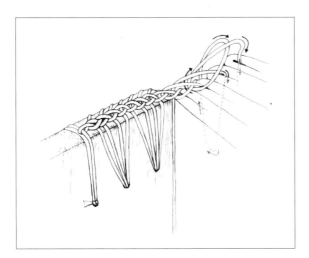

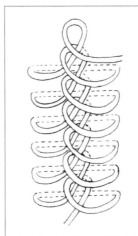

Fig. 55 (Left) Loop-stitch endband.

Fig. 56 (Right) Loop stitch in neck and armpit openings of Egyptian tunic, ca. 9th–11th century AD. Egyptian Museum of Berlin, inv. no. 9917. Redrawn and adapted from Fluck, Linscheid, and Merz, *Textilien aus Ägypten*, 200.

Twined Endbands

Twining is an ancient technique that probably developed before weaving. It has been used extensively since prehistory for starting and finishing borders not only in textiles but also in basketry and in making mats and fences for which it is still used today.[15] Although, strictly speaking, twining is different from weaving, it has also been considered a variation of weaving because it involves the presence of two sets of elements, the warp and weft, which interact. Unlike weaving, however, twining also involves interaction between the wefts.[16] In typical weaving the weft passes over and under the warps, while in twining one or more wefts also spiral (or twine) around other wefts, encircling the warps (fig. 57). Thus, in twining at least two wefts, usually of contrasting colors (fig. 58; and see fig. 64), are worked together. Although twined endbands include some of the most decorative and characteristic endbands in Eastern Mediterranean bookbinding cultures, we have no evidence of their use in Western European binding. They present the greatest variation among types of endband, and they are also the most decorative, because they allow for wide variations in colors and patterns.[17]

Fig. 57 Twining process, with two wefts (black and gray threads) intertwined around the warps (white thread).

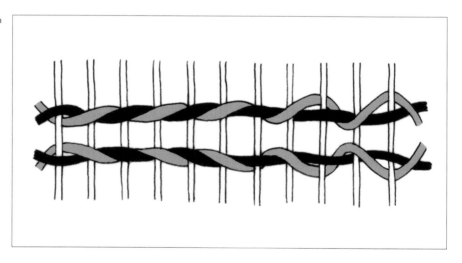

Fig. 58 Tunic fragment, Egyptian, 5th–7th century AD. Wool, linen. The red-and-white cord is partly made with twining and in this case is mainly decorative. The Metropolitan Museum of Art, Purchase by subscription, 1889, 89.18.308. Cat. 25.

Fig. 59 Tunic textile fragment, 9th century AD. Linen and wool. The red band is made with twining using threads of a single color, in this case red. The Metropolitan Museum of Art, Rogers Fund, 1966, 66.132.1.

The earliest examples of twined endbands are to be found in ninth- and tenth-century bindings from the Morgan and St. Catherine's Monastery collections. Although Petersen was familiar with the Morgan endbands, he nevertheless failed to identify them as twined endbands because in all cases they were made with plain, thick thread, which at the end of the process results in what could also appear simply to be cords sewn at the edges of the books. Among the Morgan Coptic bindings, there are at least two variations, standard twining and ply-split twining.

Standard Twining

Standard twining endbands are found in at least five of the Coptic bindings preserved at the Morgan Library that originally belonged to ninth-century manuscripts M.586 (fig. 60), 588, 590, 593, and 599. The twining twist, which alternates between the rows of twining, along with the threads used for the process, which are of the same thickness and the same natural color, results in the creation of an endband that gives the impression of a folded cord sewn on the edges of the book (fig. 60). The endbands of the tenth-century Sinai codex Georgian 18 are also formed with standard twining (fig. 61). The standard twining endband is a compound endband in which the primary component consists of overcasting sewn with thread at the head and tail edges of the codex through the centerfold of the gatherings and the spine (fig. 62). The secondary component is then made

Fig. 60 Vestige of endband from the tail edge of the right board of Morgan M.586, 9th century AD.

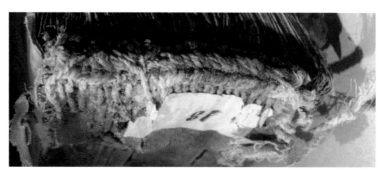

Fig. 61 Endband of Sinai codex Georgian 18, 10th century AD. St. Catherine's Monastery, Sinai, Egypt.

on top of the first, with twining using two or more threads of the same thickness and material, similar to the way twining wefts pass through the warps on the edges of textiles in antiquity and early Christian times (figs. 63 and 64). Twined endbands may have just one row of twining, as in the elewventh-century Sinai codex Georgian 17 (fig. 66),[18] two rows of usually contrasting twists, as in the Morgan bindings, or even more, as in Sinai codex Georgian 18 (see fig. 61).

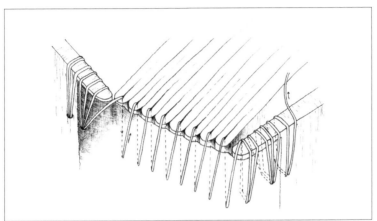

Fig. 62 Construction of the primary component of a twined endband, with overcasting through gatherings of book block. Arrows show direction of sewing.

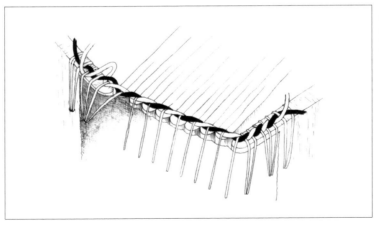

Fig. 63 Construction of twined endband. Secondary component in contrasting thread. Direction of twining is from right to left.

Fig. 64 Twined starting border from textiles, ca. 1st–6th century AD. Bergman, Nordström, and Säve-Söderbergh, *Late Nubian Textiles*, 31, fig. 27a.

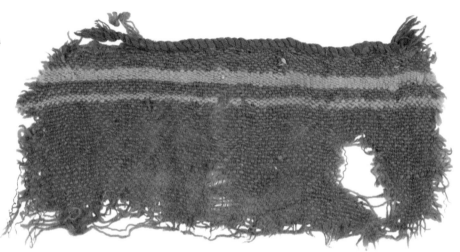

Fig. 65 Textile fragment, 5th–6th century AD, Egypt. Wool. Corded edge finish made using the green warps of the tabby weave fabric. Brooklyn Museum, Gift of the Egypt Exploration Fund, 15.475f. Cat. 24.

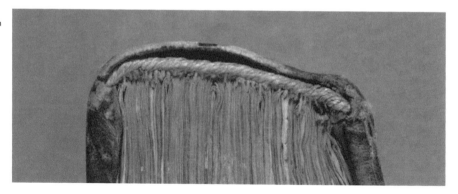

Fig. 66 Twined endband of Sinai codex Georgian 17, 11th century AD.

Ply-Split Twining

The endband vestiges on Morgan codex M.599 suggest the use of another technique, known as ply-split twining, or darning. According to Peter Collingwood, the technique, known in harness making in Egypt and Greece, is still used in present-day Turkey. The technique "can produce what looks like a warp-twined fabric but does not actually involve the active twining of warp around weft."[19] Ply-split twining is made by slightly untwisting a thread or cord, one twist at a time, to form a temporary gap between the individual strands, which then allows another thread to pass through, as in the tie-down shown in figure 67. Once this is done, the cord is left so that it returns to its former, twisted form; the process is then repeated a bit further along for the next tie-down. The vestige of the twisted cord on the detached binding shown in figure 67b also preserves what appear to be two cords tied together at the beginning of the process, with the sewing thread wound around the cords three times to secure them on the edge of the board, which would indicate the use of ply-split twining rather than standard twining. This technique may also have been used in Morgan M.588 and 590, although only a very small fragment has been preserved in M.588, and it is not possible to see the edges of the cords that might provide a definitive identification in M.590.

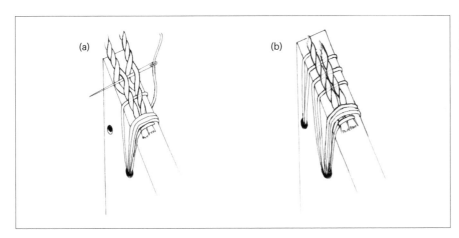

(a)

(b)

Fig. 67 Morgan M.599, 9th–10th century: (a) reconstruction of the endband-making technique; (b) surviving vestige on the binding.

Wound Endband

Although, compared to twined endbands, a wound endband looks like a rather plain endband, it is nevertheless a mechanical improvement because of the cord core around which the endband thread is wound (fig. 68). The earliest example I am aware of is found in the eighth-century Sinai codex Syriac 54, which may still preserve its original binding. From the ninth century onward, examples of this endband, one of which is to be found in Sinai codex Greek 213 (fig. 69), can be found in all Arabic, Syriac, Georgian, and Greek manuscripts in the St. Catherine's Monastery collection. The wound endband is the simplest of the endbands sewn around a core and can be found in two variations—crossed and uncrossed anchorage—depending on whether the thread crosses itself between the core and the edge of the boards and bookblock as it is wound around the core (see fig. 68). Most examples use crossed anchorage, but uncrossed examples also exist.[20] This wound endband can be directly compared to one of the commonest and simplest methods—packed overcasting—for reinforcing and decorating the edges of all sorts of fabrics, from mats and baskets to textiles (fig. 70; and see fig. 53, chap. 6).[21]

Fig. 68 Construction of wound endband.

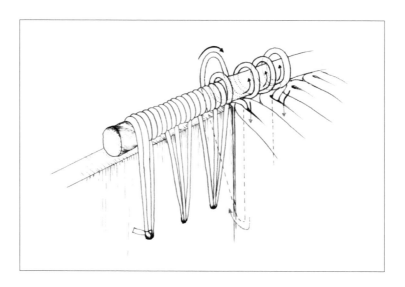

Fig. 69 The wound endband in Sinai codex Greek 213, AD 967, in contemporaneous binding. St. Catherine's Monastery, Sinai, Egypt.

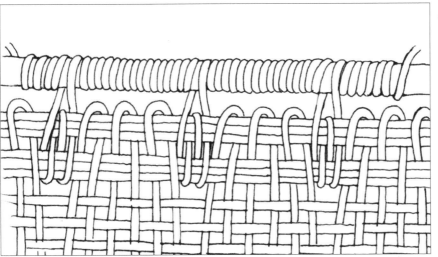

Fig. 70 Edge of curtain, Egypt, ca. 8th–10th century AD. Redrawn and adapted from Fluck, Linscheid, and Merz, *Textilien aus Ägypten*, 232.

Notes

Epigraph: Bakr al-Ishbīlī wrote *Kitāb al-taysīr fī sinā 'at al-tasfīr* between AD 1184 and 1198 in Cordoba.

1. Szirmai, *Archaeology of Medieval Bookbinding*, 23; Sharpe, "The Earliest Bindings with Wooden Board Covers," 464. Tie-downs are those lengths of the endband sewing thread that drop down from the endband at the edge of the book to the book block, pass though the gatherings, exit to the spine, and proceed upward to continue the endband sewing. Correspondingly, tie-down holes are the holes at the spine fold of a gathering through which the tie-downs pass. See also fig. 4, introduction.
2. On the Georgian bindings, see Kalligerou, "Tenth-Century Georgian."
3. Petersen, "Early Islamic Bookbindings," 54, fig. 22. See also Greenfield and Hille, *Headbands*, 10–16.
4. For a description of the technique, see Adam, "Die griechische Einbandkunst," 24; Petersen, *Coptic Bookbindings*, 35–36 and fig. 17.
5. Szirmai, *Archaeology of Medieval Bookbinding*, 21, fig. 2.3. For Chester Beatty codex 813, see Lamacraft, "Early Book-Bindings," 219, fig. 2; Petersen, "Coptic Bookbindings," 36, fig. 14a,b.
6. See Boudalis, "Endbands in Greek-Style Bindings." In his manual for bookbinding Bakr ibn Ibrahim al-Ishbili clearly states that codices bound in wooden boards should have their endbands sewn onto the boards too. See Gacek, "Arabic Bookmaking and Terminology," 109.
7. See, for example, the National Library of Greece codex 1920, dated ca. 1860.
8. See Boudalis, "The Transition from Byzantine to Post-Byzantine Bookbindings," 18–21.
9. Seiler-Baldinger, *Textiles*, 122–135.
10. "Cet ornament tressé s'exécuté d'une manière analogue à celle du 'borchmane' des burnous et djellabias" (Bel and Ricard, *Le travail de la laine à Tlemcen*). Quoted in Bosch, Carswell, and Petherbridge, *Islamic Bindings and Bookmaking*, 81, note 174.
11. See Wild, *Textiles in Archaeology*, 18. The author considers the use of these extra bundles of warp threads at the edges of textiles so characteristic of the Mediterranean—as opposed to hollow and tablet-woven selvages common in the northwest Roman provinces—that it can be used as a criterion for "identifying intrusive (imported) items in an archaeological textile assemblage of Roman date."
12. The Morgan Coptic bindings are M.574, 577, 580, 587, 597, 600, 608, 610, 634, 636, and 663bis 1. For the Sana'a bindings, see Di Bella, "An Attempt at a Reconstruction of Early Islamic Bookbinding," 107–110 and figs. 74 and 75. For the Western European bindings, see Gast, "A History of Endbands," 43–44; *Les tranchefiles brodées*, 20–23.
13. Pickwoad, "Binding," 52, 56, and fig 2.18a–c.
14. See Pritchard, *Clothing Culture*, 40–41 and fig. 3.14; Fluck, Linscheid, and Merz, *Textilien aus Ägypten*, 200.
15. See Emery, *Primary Structure of Fabrics*, 200–201.
16. Warps can be intertwined as well as wefts, for example, in tablet weaving, which was in fact used for endbands in the fifteenth and sixteenth centuries.
17. See Boudalis, "Twined Endbands."
18. Endbands with a single row of twining can be found in ten Georgian bindings from the St. Catherine's Monastery library: Sinai codices Georgian 1 and 36 (both original tenth-century bindings), Georgian 50 (tenth-century book block but rebound at a later date), Georgian 17 (eleventh-century original binding), Georgian 67 and 68 (twelfth-century book blocks, possibly in their original binding), Georgian 71, 73, 77 (thirteenth-century book blocks rebound at a later date), and 78 (eleventh-century book block rebound at a later date).
19. Collingwood, *The Techniques of Tablet Weaving*, 289–290.
20. An example of wound uncrossed anchorage is Sinai codex Greek 595 (AD 1049), in its original binding.
21. For packed overcasting, see Seiler-Baldinger, *Textiles*, 128, fig. 242a,b.

Eight

The Cover and Its Decoration

Be so good and go unto the dwelling of Athanasius, the son of Sabinus, the craftsman, and get good goat skins, either three or four, or whatsoever you shall find of good ones; and do bring them to me, that I may choose one from there for this book.

—Ostracon with letter written by Pesinthius to Peter, sixth- or seventh-century AD

W ith the advent of paperback books, the distinction between cover and boards has been lost, whereas codices can have boards without a cover—or a cover but without stiff boards. As described in chapter 5, boards are usually made of some rigid material and are connected to the front and back of a book block. The cover of a codex envelops the entire book, including the spine, and if they are included, the boards. A cover is not absolutely necessary for a bound codex—in fact, most of the earliest surviving bound multigathering codices and many later Ethiopic codices were never provided with a cover. Nevertheless, a cover has the advantage of tightening and securing the book, increasing the strength of the connection between the book block and the boards, and above all, protecting the sewing of the gatherings and the spine. As we have seen in several of the earliest preserved codices, such as the Glazier codex (Morgan MS G.67), a leather cover was provided only for the spine of the bound codices and was also used for attaching the boards. A cover had to be flexible, strong, and suitable for different types of decoration. The material that best combines all these characteristics is tanned leather.

The process for creating a cover is quite simple: a piece of leather is pasted on the outer face of the spine and boards of the codex and then folded over the outer edges to the inside. Leather can take various types of decoration, all well known from late antique and earlier artifacts of the same geographical and cultural context, such as sandals, shoes, and belts. All decorative techniques found in early bookbindings can also be found in shoes.[1] The similarity of the most elaborate decorated covers to the ornamentation of shoes is so striking that it seems reasonable to suggest that both were done by the same artisans. Embroidery, the only technique that is not found among the surviving bindings, is not very common in shoe decoration either.[2]

Book covers made of luxury materials, such as ivory and precious metals, and occasionally gems and pearls, were probably fashioned by goldsmiths or other craftsmen (fig. 71). They were not integral to the structure of the book but were instead constructed separately and then nailed onto the boards. Such covers are evidence of the centrality of books for Christians in their beliefs and rituals, through which a book could turn from a text container into a precious and sacred object full of symbolic meaning. Books held by Christ or the saints are extensively represented in iconography and occasionally drew criticism for their extravagance.[3] In a letter to Eustochium, written in AD 384 in Rome, St. Jerome wrote: "The more scrupulous sort wear one dress till it is threadbare, but though they go about in rags their boxes are full of clothes. Parchments are dyed purple, gold is melted for lettering, manuscripts are decked with jewels: and Christ lies at their door naked and dying."[4]

Most bindings were decorated with a combination of techniques, which will be considered separately in the remainder of this chapter.

Fig. 71 Cover of the Gospels of Theodolinda, Queen of Lombardy, possibly made in Rome, late 6th–early 7th century AD. Gold, precious stones, enamels, and cameo on a wooden core. Donated by the queen to the Basilica of John the Baptist, Monza, Italy. Monza Cathedral Treasury, Italy.

Tooling

Tooling is the decoration of leather by pressing or hammering the surface with metal stamps (or "tools"), usually with intaglio decorative motifs, leaving a permanent relief impression on the leather (fig. 72).[5] In other artifacts, such as shoes, the process is also known as impressing or stamping (fig. 73).[6] The leather is usually damp, and the tool may or may not be heated before it is pressed on the leather surface.[7] The same process, already used in the Nag Hammadi codices, would normally start with creating a grid to define the space to be decorated. Single or multiple lines are made on the leather with a pointed but usually blunt tool, such as a bone folder, stylus, or metal "creaser." Once the decorative grid was outlined, metal tools—each bearing a single motif—were used (figs. 73 and 74). Both Hobson and Petersen compiled lists of tools from a total of sixty-nine decorated bindings from the sixth to tenth centuries in the Morgan and from European and Egyptian collections. Altogether, they present a rather limited variety of motifs, most of them simple, geometric, or floral, with a few figurative motifs, such as birds, animals, and a human figure riding a horse.[8]

Fig. 72 Tooling on Sinai codex Georgian 32 (detail), AD 864. See also fig. 75, in which belts 695 (far right), 696 (top right), and 697 (far left) would have been decorated with a similar X-shaped tool.

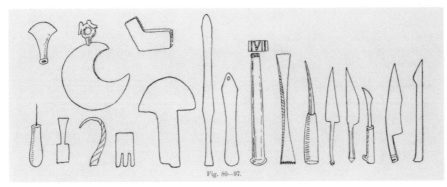

Fig. 80—97.

Fig. 73 Shoemaking tools, Akhmin-Panopolis, Egypt, late antiquity. The eighth tool from the right is for decorating leather. From Frauberger, *Die Antiken und frühmittelalterlichen Fussbekleidungen aus Achmim- Panopolis*, 48, fig. 80–97. Göttingen State and University Library.

Fig. 74 Left: tool for the decoration of bindings, St. Catherine's Monastery, Sinai; middle: the intaglio surface of the tool; right: smoke impression of tool on paper. From Sarris, *Classification of Finishing Tools in Greek Bookbinding* 1:63, fig. 24.

Of particular interest are small motifs consisting of two or three concentric rings, commonly found, for instance, in sandals.[9] These very simple motifs are ubiquitous in Byzantine bindings until as late as the seventeenth century.[10] Concentric rings were also used in all sorts of different artifacts from the early Christian period—from amulets and combs to mosaics and architectural elements—and appear to have had a strong apotropaic function, suggesting eyes or mirrors, which since classical antiquity were believed to have protective power (figs. 75, 76, 77, and 78; see also fig. 85; figs. 107, 108, 109, chap. 9; fig. 115, chap. 10).[11] Besides the concentric rings, there are other, mostly simple tooled motifs, such as +, X, and M designs, that can be found on codex covers and also other leather artifacts, such as belts (see figs. 72 and 75).

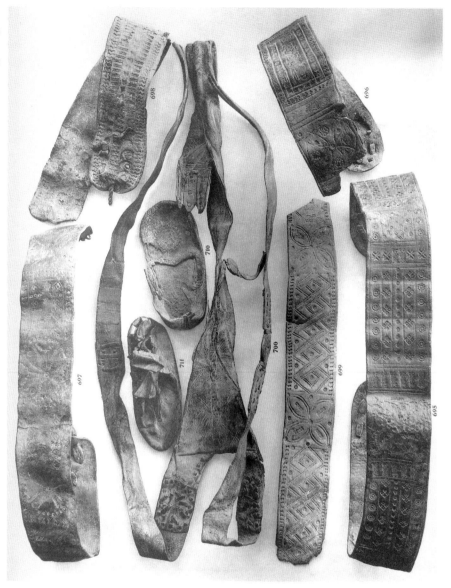

Fig. 75 Coptic leatherwork. Belts (nos. 695–699), *analabos* (no. 700), and shoes (nos. 710 and 711), From Abu Tig, Luxor, Akhmin, and Giza. From Wulff, *Altchristliche und mittelalterliche byzantinische und italienische Bildwerke,* fig. 31.

Fig. 76 Coptic leatherwork. Book cover fragments (nos. 693 and 694), coiffure supports (nos. 702 and 703), shoes (nos. 708, 709, and 712), and possibly bookmarks (689, 690). From Abu Tig, Cairo, Giza, and Akhmin-Panoplois. From Wulff, *Altchristliche und mittelalterliche byzantinische und italienische Bildwerke,* fig. 30.

Fig. 77 Lower cover of Morgan M.569, 9th–10th century AD. Redrawn and adapted from drawing by Theodore C. Petersen, ca. 1940.

Fig. 78 Upper cover of Morgan M.569, 9th–10th century AD. Leather over papyrus board; parchment, thread. Decorated with cutout openwork, stitching, gilding, and lacing. The Morgan Library and Museum, purchased for J. Pierpont Morgan, 1911. MS M.569A. Cat. 33

Sometimes the force needed to impress the tools on the leather could be considerable. For some bindings—especially those bound in papyrus laminate boards—the leather may have been decorated separately before the cover was adhered to the boards of the bound codex. Bindings with double boards (discussed in chapter 5), also common in this period, offered another way to handle the force required for decorative tooling without damaging the book block (see fig. 51, chap. 5).

Inked and Painted Decoration

Using paint or ink to produce motifs on the leather of codex covers appears to have been fairly common from late antiquity until the tenth century. We have an almost continuous sequence of examples: the binding of Nag Hammadi codex II is the earliest, followed by the bindings of Chester Beatty codices 813, 814, and 815 (see fig. 92), the British Library Greek Papyrus 1442 (eighth century), and Berlin Egyptian Museum P. 14016 (ca. eighth–tenth century).[12] To these well-known examples, a few outstanding examples of Georgian codices from the St. Catherine's Monastery library should be added—for example, Sinai codices Georgian 31, 37, and 63, all written in the tenth century and still preserving their original bindings. In most of these bindings, ink appears to have been used to draw the decorative pattern, which was subsequently tooled or incised (fig. 79). Ink was also used to draw dots or other patterns on parchment patches that backed cutout patterns in several Coptic bindings from the Morgan collection.[13]

Boards or covers painted with opaque pigment are rare, but we do have early examples even among wooden tablet codices, such as the cover of a group of tablets from Egypt, first covered with a thin layer of stucco and then painted with the three forms of Thoth.[14] The best-known examples, however, are the two boards from the Freer Gospel (Freer Gallery of Art, F1906.297), which are supposed to have been painted on the wooden boards of an early fifth-century Greek Gospel codex sometime in the first half of the seventh century.[15] There is also a similar decoration on a small fragment that appears to have belonged to the leather cover of a codex bound between the fifth and seventh centuries, originally decorated with two figures holding supposedly sacred books.[16] Examples with nonfigurative decoration painted with opaque colors are Berlin

Egyptian Museum P. 14016 (ca. eighth–tenth century) and Morgan M.601 (ninth–tenth century), which is decorated with a geometrical pattern painted in yellow.[17] To the best of my knowledge, there are no examples of shoes decorated with ink, although dark dyes have been occasionally used (see fig. 88), and the body color would have been inadequate for decorating shoes because the pigment would have peeled off before long.

Of course, one should also mention the option of dyeing the leather itself, although all surviving examples seem to indicate that the selection of colors for the leathers used for book covers was usually restricted to hues of brown, red, and black.

Fig. 79 Lower cover of Sinai codex Georgian 31, 10th century AD. Interlaced pattern drawn with ink, then incised with sharp tool. The star-shaped patterns are tooled.

Cutout Openwork and Stitching

Trimming the edges of leather to produce decorative shapes is a technique that was used in Roman and late antique shoes.[18] In bookbinding, the only examples we have are the leather spine cover extension of the Glazier codex (Morgan MS G.67, see figs. 45 and 46, chap. 5) and some similar finishes represented in various images of books from late antiquity, in which the head and tail edges of the spine are often shown as having foliated extensions, often with three lobes (see, for example, the tripartite extension from the head edge of the open codex held by Saint Luke in figure 42, chap. 5). The technique is also used for bookmarks, as we will see in chapter 10. Another technique, cutting out motifs on the leather before it is used to cover a book and superimposing it over other material to form decorative patterns and effects, is also well documented in Egypt from at least the New Kingdom and Third Intermediate Period (ca. 1550–712 BC).[19] There are two variations of this technique, both extensively used in late antique shoes from Egypt as well as in some bookbindings.

The first variation consists of sewing the upper cutout layer on top of another layer so that the layer below will show through. The lower layer is parchment or leather and is often gilded. There are superb examples of shoes produced with this very elaborate technique (fig. 80; see also figs. 76 and 89), as well as a number of peculiar coiffure supports from Egypt dated to the third–seventh centuries AD (fig. 81; see also items 702 and 703 in fig. 76).[20] There are four bindings dated before the tenth century decorated with this technique, which also possibly represent the most elaborate examples of decorated leather bindings from this early period. The first example is Morgan codex cover M.569, ca. 9th–10th century AD (see figs. 77 and 78);[21] the second, codex cover P. Vindob. G 30501 of the Austrian National Library, ca. eighth–ninth century AD (fig. 82);[22] the third, Egyptian Museum Berlin P. 14018, ca. eighth–ninth century (see fig. 95).[23] The fourth example is the eighth-century leather binding fragment of unknown provenance at the British Museum (EA67080); although badly preserved, it is clear due to the partly preserved inscriptions on its surface, that it is decorated with the effigies of Mathew, Emmanuel (Christ), Mark, Luke, and John, indicating

Fig. 80 Leather shoe appendages decorated with cutout openwork and stitching, late antiquity (ca. AD 300–700). Leather. Courtesy of Penn Museum, 2003.34.351L and 2003.34.315BC. Cats. 38 and 39.

Fig. 81 Coiffure support decorated with cutout openwork and stitching, Akhmim-Panopolis, ca. 3rd–7th century AD. Leather, textile, thread. The Metropolitan Museum of Art, Gift of George F. Baker, 1890, 90.5.40. Cat. 37.

that it once covered a manuscript of the four Gospels (fig. 83).[24] In all these bindings except the Berlin cover, the lower layer is gilded, producing an impressive effect. The gilded parchment or leather layers are not a single uniform piece but rather are constructed of several irregular pieces (or scraps), which suggests that despite their luxury and high craftsmanship, these bindings were made in workshops where nothing was too small or worthless to be used.[25] The top cutout patterns are sewn on the gilded backgrounds with thin thread.

In the second variation, parchment backs only the cutout areas of the leather and is just pasted, rather than sewn.[26] As many as nineteen codices from the Morgan collection have bindings decorated with this variation, which usually consists of simple geometric shapes cut out through the leather cover, which is backed with pieces of parchment that are only big enough to back the cutout pattern. The parchment is often colored yellow, possibly in an attempt to imitate gold (fig. 84). In some cases, the parchment that shows through the cutouts is further decorated with simple drawn patterns, such as dots and circles. The same technique is also used in the fragment of the binding from Manichaean codex MIK III 6268 (eighth–ninth

Fig. 82 Theodore C. Petersen. Facsimile of a Coptic binding pattern from a cover in the Austrian National Library, Vienna, (P. Vindob. G 30501, ca. 8th or 9th century AD) ca. 1940. The Morgan Library and Museum, PCC 93. Cat. 29.

Fig. 83 Fragment of a bookbinding (and drawing) decorated with the effigies of Christ Emanuel and the four Evangelists using the same techniques as in the bindings shown in figures 77, 78, and 82. The strips of leather stitched on the gilded background are almost completely lost leaving the sewing holes. Tanned leather, gilded leather, linen, thread. Egypt, 8th century AD. © Trustees of the British Museum EA 67080.

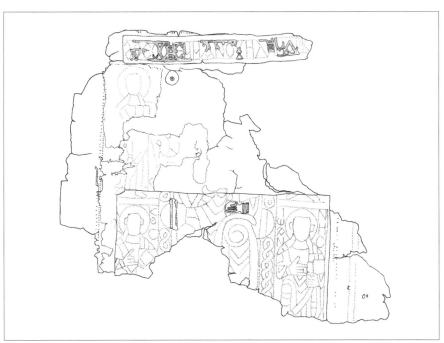

century) found in Turfan and now preserved in the Museum of Indian Art in Berlin and in the original eighth-century cover of the Ragyndrudis codex, which was written and presumably bound in Luxeuil, France.[27]

Finally, it should be noted that in specific areas of the cover there could be more than just two layers; for example, the binding of Morgan M.569 contains at least four different layers of leather in the small roundels arranged around the central crosses on the cover of the two boards (see figs. 77 and 78).

Incising and Scraping of the Surface

As mentioned earlier in this chapter, in some tenth-century Georgian bindings at St. Catherine's Monastery, the patterns of the cover decoration are first drawn with ink and then incised with a sharp tool, possibly a knife (fig. 85; see also fig. 79).[28] A similar technique consists of scraping the surface of the leather with a sharp tool so that the underlying leather, which is fibrous and a lighter color, is exposed, creating a simple contrasting decorative effect. Examples of this method can be found in bindings in the Morgan collection (for instance, M.577 [fig. 86]), as well as in other collections. Scraping can also be found in shoes and sandals (figs. 87 and 114).[29]

Fig. 85 Decorative tooling (detail), Sinai codex Georgian 32, AD 864. St. Catherine's Monastery, Egypt.

Fig. 86 Upper cover originally on MS M.577, Egypt, 9th–10th century AD. The Morgan Library and Museum, Purchased for J. Pierpont Morgan, 1911, MS M.577A1. Cat. 21.

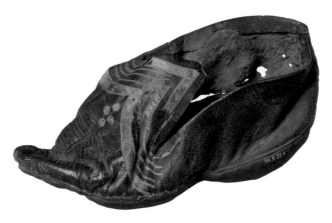

Fig. 87 Slipper, Egypt, 3rd–7th century AD. Leather. The Metropolitan Museum of Art, Gift of George F. Baker, 1850, 90.5.37a. Cat. 36.

Gilding

To judge from surviving Eastern Mediterranean examples from late antiquity, gilding in leatherbound books was rather rare. When gold ornamentation was desirable and affordable, it could also take the form of covers made of precious metals such as those seen in figure 71. Leather gilding is found in the eighth- or ninth-century bindings of Morgan M.569 (see figs. 77 and 78), Vienna P. Vindob. G 30501 (see fig. 82), and British Museum EA 67080 (see fig. 83); in the spine of Chester Beatty codex 814; and in vestiges of the original tenth-century binding with double boards of the Sinai codex Greek 213. In the Morgan M.569 and the Vienna bindings, the visible gilding is from the gilded parchment sewn under the cutout leather cover. Leather gilding and its different techniques are well represented in shoes from the same period and was employed to define and form patterns and motifs (fig. 88), as an underlying layer showing through cutout patterns (fig. 89), or even in the form of fully gilded shoes.[30]

Fig. 88 Pair of shoes, Egypt, 5th–8th century AD. Leather with gilding and dyeing. Byzantine and Christian Museum, Athens, BXM 00507.

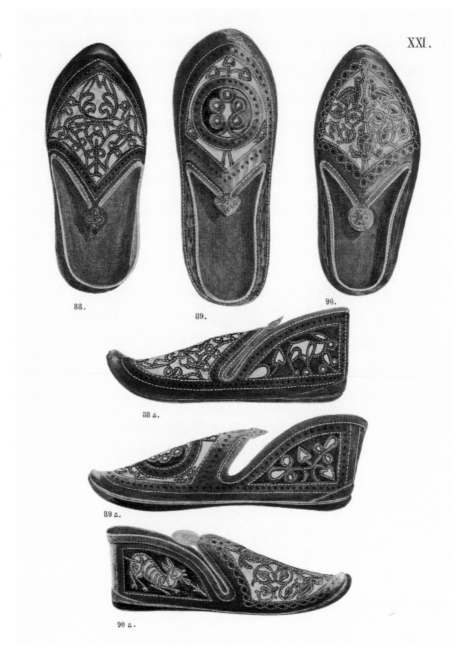

Lacing

The term "lacing" refers to the weaving of thin parchment strips through closely spaced slits cut through the leather, alternating between front and back, thus creating a simple but distinctive decorative effect.[31] To give but one prestigious example, lacing, which has been known since antiquity, is used in the decoration of King Tutankhamun's sandals.[32] The technique is also well documented in shoes from late antiquity (fig. 90). In bookbindings, lacing can be found in several codices such as Leyden Anastasy 9 (fifth century); the Morgan fragment M.614bis (seventh or eighth century); Vienna P. Vindob. G 30501 (eighth or ninth century; see fig. 82); Cairo Egyptian Museum Hamuli J. 47559 (ninth century); Morgan M.569 (late ninth or early tenth century; see figs. 77 and 78); and in some undated fragments in the Chester Beatty Library (Cpt. 808).[33]

Fig. 90 Pair of shoes, Egypt, ca. 3rd–9th century AD. Leather with cutout openwork, stitching, gilding, tooling, and lacing. The Metropolitan Museum of Art, Gift of Edward S. Harkness, 1926, 26.9.11a,b. Cat. 35.

The Patterns

In his guide to divine and secular literature, written around AD 550, Cassiodorus alludes to collections of decorative patterns from which a bookbinder could choose:

> We have provided workers skilled in bookbinding, in order that a handsome external form may clothe the beauty of sacred letters. . . . And for the binders, in a fitting manner, unless I err, we have represented various styles of bindings in a single codex, that he who so desires may choose for himself the type of cover he prefers.[34]

His statement does not seem at all improbable, given that patterns used by Egyptian weavers from the sixth century and earlier have been preserved.[35] Decorative patterns have always been quite popular among art historians for studying and dating bookbindings.[36] Interlaced patterns—in which bands are passed in front of and behind other bands to create a complex decorative design—are by far the most common patterns for bindings covered in leather during late antiquity. Although they seem to have been uncommon in Byzantine bindings, surviving examples suggest that interlaced patterns continued to be used in Islamic, Georgian, and Armenian bindings after the tenth century, and in Western European bookbinding from the Renaissance onward.

James Trilling separates interlaced patterns into two categories: medallion and nonmedallion interlaces.[37] Medallion patterns are by far the more common in late-antique bindings (see, for example, figs. 77, 78, and 82; also figs. 97 and 98). Nonmedallion patterns are also called complex interlaces (see figs. 79 and 93). In the great majority of bindings before the tenth century, the oblong surface of the cover is divided to form a square space in the center, usually framed by two decorative bands at the top and bottom edges and one to the left and right. This arrangement of the decorative space can be found in manuscript bindings and illuminations (see figs. 77, 78, and 82) and is also very commonly used in mosaic floors of late antiquity (fig. 91). In later bindings, in both the West and the East, the central panel is oblong, following the shape of the boards themselves; thus, the number

Fig. 91 Mosaic pavement,
Capua, Italy, pre–9th century
AD. Redrawn from Garrucci,
Storia dell'arte cristiana,
table 277.

of decorative bands surrounding the central oblong panel does not vary as it does when the central panel is square.[38]

Although there are endless variations, most interlaced patterns in bookbindings can be regarded as elaborations on an underlying shape—a rectangle, a lozenge, or a circle—inscribed inside the cover's decorative field. Several of the Morgan Coptic bindings are decorated with lozenges or with two interlaced rectangular shapes (see, for instance, fig. 86, Morgan M.577, in which a central square panel is placed within a rectangular border).[39] The decorative field can also be filled with diagonals that intersect to create linear interlaces, which Trilling calls a "grid-plait."[40] The earliest preserved example of such a pattern in a binding is the inked decoration of the small sixth-century Chester Beatty codex 815 (fig. 92). The same grid-plait pattern can be found among tenth-century Georgian bindings from St. Catherine's Monastery, as well as extensively in Kairouan bindings and later in Armenian bindings.[41]

Patterns built on a circle usually incorporate more or less complex motifs, for instance, two squares interlaced inside a circle to create an eight-pointed star.[42] The earliest example of this pattern appears in a grafitto from the first century AD that was excavated at Dura-Europos.[43] It is very common among textiles from the third through sixth centuries

Fig. 92 Chester Beatty codex 815 (6th century AD). From Lamacraft, "Early Book-Bindings from a Coptic Monastery," fig. 8.

(fig. 94) and in mosaics (as seen at the center of fig. 91), stelae, architectural elements, from plans of entire buildings to small objects, such as lamps and shoes (see figs. 80 and 88), and in codex illuminations as well as covers (figs. 95 and 96). Ranging from Egypt to Syria and from Carthage and Split to Constantinople, the pattern appears to have had cosmic, Christian, and imperial significance.[44] This basic pattern can be further elaborated to produce some of the most intricate examples of interlaced patterns we have in surviving bindings, such as those of Morgan M.569 (see figs. 77 and 78), Vienna P. Vindb. G. 30501 (see fig. 82), and Ambrosianus Syrus C. 313 inf (fig. 97) codices.[45]

A four-leaf interlaced pattern is often inscribed in the central square field of book covers and can be further interlaced through a circle (fig. 98).[46] In its simplest form, this pattern has a pre-Christian history, but with the circle it appears to be specifically connected with Christianity. According to Patrick Reuterswärd, the pattern was "usually constructed with a compass" and consisted of "four centrally facing semicircles, the sides of which intersect.

Fig. 95 Theodore C. Petersen. Facsimile of a Coptic binding pattern from a cover in the Egyptian Museum Berlin (P. 14018, ca. 8th–9th century AD), ca. 1940. The Morgan Library and Museum, PCC094. Cat. 30.

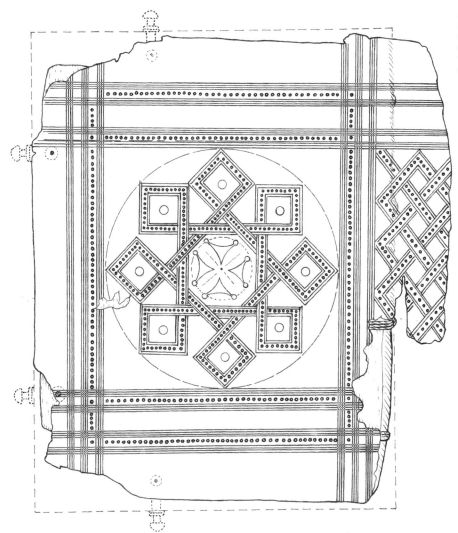

Fig. 96 Theodore C. Petersen.
Pattern of lower board and
cover, Morgan codex M.574,
(9th–10th century AD), ca.
1940. The Morgan Library
and Museum, PCC059.

Fig. 97 Complex medallion interlaced pattern tooled on lower board cover (this page) and upper board cover (opposite page), Syriac codex Ambrosianus Syriacus C 313 inf. Each board approx. 13⅜ × 10⅝ in. (34 × 27 cm). Boards and leather cover originally made in the 6th–7th century for a two-volume Greek Gospel, later reused to bind the current Syriac codex. Title written in Greek on the two boards reads: "First Book of the Sacred / Scriptures of the Old and / New Testament Belongs to / Basilios the Deacon." Redrawn and adapted from Petersen, "Coptic Bookbindings," fig. 20, after examining the original covers.

To this basic figure, with its four diagonally protruding pointed ellipses, may be added concentric circles, of which one should intersect the ellipses. By the addition of such a circle, the interstices within take on the shape of a cross."[47] The pattern and its variations are very common in Roman, Coptic, Byzantine, Celtic, Armenian, and Lombard art, usually found with the four-looped pattern in a resting position (X) or, more rarely, in a standing position (+). In all cases, as Reuterswärd notes, the pattern is "intended as an alterna-

Fig. 99 Cover decoration based on an elaboration of the four-leaf interlaced pattern shown in figure 98, creating a central cross pattern. Sinai codex Georgian 32, written and presumably bound in AD 864.

tive to the symbol of the Cross."[48] With slight modifications, it can form an interlaced cross, which is common on the covers of a number of Georgian codices, such as Sinai codices Georgian 32 (AD 864) (fig. 99) and Georgian 47 (AD 977). In Georgian 47 the same motif is found inside the manuscript as a full-page miniature (folio 81) as well as on its cover. In Morgan M.600, the four-leaf motif is interlaced through a square rather than a circle. Again, these are all patterns used in late antique floor mosaic (fig. 100).

Despite St. John Chrysostom's insistent warnings and attacks against amulets, incantations, and any other "unofficial" protective devices, the use of apotropaic (to turn away evil) symbols other than the cross was extensive.[49] The Gospel itself and other sacred books were used as amulets and even to cure headaches or fever, a practice that both St. Augustine and St. Jerome disapproved of.[50] There are reasons, then, to believe that at least some of the patterns and motifs used for the covers of bound codices had more than a decorative role. Interlaced motifs, along with the cross and images of Christ and saints, may have had a protective and apotropaic role. Motifs that are intricate and complex elaborations of knots have been widely used since antiquity not only as apotropaic devices and good luck tokens but also for performing magic and "binding" spells.[51] Knots, interlaces, and crosses are often combined in early Christian mosaics as well as in doors and thresholds as apotropaic devices.[52] As Trilling notes, "manuscripts, too, have their 'doorways'—frontispieces, final pages—and these too require protection."[53] Crosses have been a constant motif in Christian art, ubiquitous in all contexts and materials as a protective symbol, device, or gesture. In books, they have been widely used on the outside of covers of bound codices, inside in the form of illuminations on the pages of manuscripts, especially in the frontispieces of Insular (Irish and British), Coptic, and Byzantine codices, and also in Armenian manuscripts, in which crosses are commonly combined with interlaced patterns, and even as amulets concealed in spine linings, as noted in chapter 6.

Notes

Epigraph: *P.Mon.Epiph.* 380. From Crum and White, *The Monastery of Epiphanius at Thebes*, pt. 2, 254. Translation from Kotsifou, "Books and Book Production," 61.

1. On shoes in Greco-Roman antiquity, see Knötzele, *Auf Schusters Rappen*; Forbes, "Footwear in Classical Antiquity," 58–63. See also Frauberger, *Antike und frühmittelalterliche Fussbekleidungen aus Achmim-Panopolis.*

2. For examples of embroidered shoes, see Frauberger, *Antike und frühmittelalterliche Fussbekleidungen*, cat. nos. 23, 56, 62, 68, 83, 93. See also Grohmann and Arnold, *Denkmäler islamischer Buchkunst*, 17b.

3. For covers made with precious metals, see Lowden, "The Word Made Visible," 31–44.

4. Quoted in Saint Jerome, *Select Letters of St. Jerome*, 32.

5. In late antiquity and as late as the Renaissance, the motifs of tools were engraved by hand, while in later periods they were cast or both cast and engraved. See Sarris, "Classification of Finishing Tools in Greek Bookbinding," 1:103–104. See also Szirmai, *Archaeology of Medieval Bookbinding*, 242–245.

6. For impressing and stamping, see Veldmeijer, *Sandals, Shoes and Other Leatherwork*, 25–26. See also Van Driel-Murray, "Leatherwork and Skin Products," 307.

7. According to Adam, "Die griechische Ein-
    bandkunst," and Ibscher, "Von der Papyrus-
    rolle zum Kodex," the tools used in these
    bindings were pressed unheated. See also
    Petersen, "Coptic Bookbindings," 76–77; Sar-
    ris, "Classification of Finishing Tools," 1:106;
    Szirmai, *Archaeology of Medieval Bookbind-
    ing*, 246.
8. See Hobson, "Some Early Bindings and
    Binders' Tools," fig. 2; Petersen, "Coptic
    Bookbindings," figs. 42a,b. For a selection
    of very similar tools (from a total of about
    240 different examples) used in the Kairouan
    bindings, see Petersen, "Early Islamic Book-
    bindings," 47, fig. 20.
9. Double or triple concentric ring motifs are
    found, for example, in sixteen sandals (cat.
    nos. 9, 23, 24, 29, 30, 31, 32, 33, 34, 35,
    36, 37, 38, 39, 41, 51, and 53), most dated
    between AD 25 and 850 (only cat. nos. 31,
    33, and 53 are dated after 1100) (Veldmeijer
    and Ikram, *Catalogue of the Footwear in the
    Coptic Museum*); in four sandals of unknown
    dates from the Louvre collection (Montem-
    bault and Musée du Louvre, *Catalogue des
    chaussures*; cat. nos. 43, 50, 53); and in
    similar artifacts in several other collections
    (there are too many to list in detail).
10. On concentric ring motifs in Byzantine and
    post-Byzantine bookbindings, see Boudalis,
    "The Transition from Byzantine to Post-
    Byzantine Bookbindings," 27 and figs. 29a,b.
11. On the use of concentric rings as apotro-
    paic devices in mosaics and architectural
    elements, see Maguire, Maguire, and Dun-
    can-Flowers, *Art and Holy Powers in the Early
    Christian House*, 5–7; Maguire, "Magic and
    Geometry in Early Christian Floor Mosaics
    and Textiles," 267; Wulff, *Altchristliche und
    mittelalterliche byzantinische und italienische
    Bildwerke*, plates 10, 11, 24, 31.
12. The patterns in the covers of these three
    multigathering codices are reproduced in
    Petersen, "Early Islamic Bookbindings,"
    55–57, figs. 23, 24, 26.
13. Ink patterns on parchment patches can be
    found in Morgan codices M.570, 574, 575,
    576, 577, 580, 582, 583, 585, 586, 588, 590,
    593, 594, 595, 599, 600, 604, and 663bis4.

14. For a description, see Petrie, *Objects of Daily
    Use*, 66, cat. no. 65, n.d.
15. For a recent discussion of the two covers,
    see Lowden, "The Word Made Visible," 21–23.
    See also Morey, "The Painted Covers."
16. See Del Corso and Pintaudi, "Papiri letterari
    del Museo Egizio del Cairo."
17. For the Berlin codex, see Petersen, "Early
    Islamic Bookbindings," 56, fig. 26. For the
    Morgan codex, see Depuydt, *Catalogue of
    Coptic Manuscripts in the Pierpont Morgan
    Library*, 62–64; Petersen, "Coptic Bookbind-
    ings," 216–218 and fig. 42.
18. For examples of decorative trimming in
    shoes and sandals, see Frauberger, *Antike
    und frühmittelalterliche Fussbekleidungen*,
    cat. nos. 14, 15. See also Van Driel-Murray,
    "Vindolanda and the Dating of Roman Foot-
    wear," esp. fig. 1. Also Waterer, "Leatherwork,"
    179–193.
19. For the processing and use of leather in
    ancient Egypt, see Van Driel-Murray, "Leath-
    erwork and Skin Products," 309–312.
20. For shoes, see Montembault and Musée du
    Louvre, *Catalogue des chaussures*, 57–59,
    cat. nos. 80, 81, 125; Veldmeijer and Ikram,
    Catalogue of the Footwear, cat. nos. 61, 62,
    64, all of them dated as Christian or Coptic;
    Frauberger, *Antike und frühmittelalterliche
    Fussbekleidungen*, cat. nos. 20, 31, 32, 33,
    39, 75, 76, 77, 78, 79, 80, 85, 86, 88, 89, 90,
    95. For head supports, see Wulff, *Altchristli-
    che und mittelalterliche byzantinische und
    italienische Bildwerke*, 159, plate 30, cat. no.
    703; Strzygowski, *Koptische Kunst*, 165, plate
    10, cat. no. 7246. The coiffure supports in the
    Metropolitan Museum of Arts are acc. nos.
    90.5.39 and 90.5.40, both dated between
    the fourth and seventh centuries AD.
21. For the Morgan M.569 cover, see Needham,
    Twelve Centuries of Bookbindings, cat. nos.
    2, 13–16. Petersen suggests that the cover
    might have been salvaged from an earlier
    binding and reused ("Coptic Bookbindings,"
    100–106).
22. On P. Vindob. G 30501, see Arnold and
    Grohmann, *The Islamic Book*, 34, 35, and
    fig. 16. Also Ruprechtsberger, *Syrien*, 485
    and 432, pap. 32.

23. On Berlin Egyptian Museum P. 14018, see Schubart, *Das Buch bei den Griechen und Römern*, 142.

24. On British Museum EA 67080, see Shore, "Fragment of a Decorated Leather Binding from Egypt." About the decoration of covers with effigies, see Kotsifou, "Bookbinding and Manuscript Illumination," 226–227.

25. Although not described in the text, the illustration provided suggests that a similar practice might have been used in a very elaborate shoe from Egypt. See Frauberger, *Antike und frühmittelalterliche Fussbekleidungen*, cat. no. 88; reproduced in fig. 85.

26. For the use of this technique in shoes, see Montembault, *Catalogue des chaussures*, cat. nos. 55, 59, 60, 65; Veldmeijer and Ikram, *Catalogue of Footwear*, cat. nos. 132, 139, all identified as Christian or Coptic. See also Frauberger, *Antike und frühmittelalterliche Fussbekleidungen*, cat. nos. 30–33, 36, 37, 75–80, 84, 85–87, 88–90, 95.

27. On the Manichaean codex MIK III 6268, see Gulácsi, *Medieval Manichaean Book Art*, 83–88 and figs. 3/12 and 3/14. On the Ragyndrudis codex, see Szirmai, *Archaeology of Medieval Bookbinding*, 97.

28. Examples are Sinai codices Georgian 31 (Epistles and Acts, tenth century) and 70 (Triodion, thirteenth century).

29. Morgan examples include M.570, 577, 581, 600, 604; other examples are British Library Or. Ms. 3367 and Berlin Egyptian Museum P. 14019; for shoes, see Frauberger, *Antike und frühmittelalterliche Fussbekleidungen*, cat. nos. 18, 19, 25, 26, 27–29, 30, 31, 44, 94; Veldmeijer and Ikram, *Catalogue of Footwear*, cat. no. 130.

30. For gilding in shoes, see Frauberger, *Antike und frühmittelalterliche Fussbekleidungen*, cat. nos. 16, 17, 30, 34–36, 60, 61, 63b, 67, 68, 69–73, and 87. The Abegg-Stiftung collection contains a pair of fully gilded leather shoes from the sixth or seventh century (see De Moor et al., "Radiocarbon Dating and Colour Patterns," 176), and a single shoe, fourth–seventh century, from the DLM Deutsches Ledermuseum / Schuhmuseum Offenbach (Göpfrich, Frankenhauser, and Mackert,

Wettlauf mit der Vergänglichkeit, 131–134, cat. no. 15). Both are also decorated with what is described as embossing, or patterns formed through repetitive and closely spaced dot impressions on the leather. For gilding in shoes in general, see Nauerth, "Sandalen, Schuhe und Pantoffeln mit Vergoldung."

31. Petersen calls this decoration technique "braid work" ("Coptic Bookbindings," 83).

32. See Veldmeijer et al., *Tutankhamun's Footwear*, 162, fig. 415.

33. For the Leyden binding, see Petersen, "Early Islamic Bookbindings," fig. 33; for the Cairo binding, see Petersen, "Coptic Bookbindings," pt. 3, item 89. Grohmann and Arnold also mention an eighth- to ninth-century binding with lacing at the Egyptian Museum of Berlin (*Denkmäler Islamischer Buchkunst*, 18).

34. Quoted in Scheller, *Exemplum*, 19.

35. See ibid., 94–97, cat. no. 2. Also Stauffer, "Cartoons for Weavers from Greco-Roman Egypt."

36. The bibliography on decorative patterns used in bookbinding covers is very extensive. For works related mostly to the Eastern Mediterranean book structures, see Hobson, "Some Early Bindings and Binders' Tools"; Ettinghausen, "Near Eastern Book Covers"; Federici and Houlis, *Legature bizantine vaticane*; Boudalis, "The Evolution of a Craft"; Sarris, *Classification of Finishing Tools*.

37. See Trilling, "Medieval Interlace Ornament", 61. See also Åberg, *The Occident and the Orient*, 61–77, 118–121.

38. For an analysis of decorative patterns in which the number of decorative bands varies between the top and bottom and the left and right sides of the central panel, see Stevick, "St. Cuthbert Gospel Binding." For an example from an illuminated manuscript, see folio 192 verso of the Book of Durrow, from the second half of the seventh century. For a categorization of the decorative patterns in Eastern Mediterranean bookbindings, see Boudalis, "The Evolution of a Craft," 811–813; Sarris, *Classification of Finishing Tools*, app. 5, 124–129.

39. The following are bindings decorated with a lozenge from the Morgan collection: M.577,

581, 582, 583, 586, 594, 595, 599, 603, 603bis, 609, 614bis, 633, 634, and 660. Examples from other collections are Berlin Egyptian Museum P. 1402, Cairo Coptic Museum Hamuli E and Hamuli G, and British Museum Oriental 7597 and 6801. The following are bindings decorated with two interlaced rectangles: Morgan M.581 and 663bis, BML Pap V, and Cairo Coptic Museum Hamuli B.

40. Trilling, "Medieval Interlace Ornament," 73.

41. Examples of grid-plait patterns in bindings from Kairouan bindings are no. 72 of the tenth century and no. 98 of the eleventh century (Petersen, "Early Islamic Bookbindings," figs. 16 and 17).

42. Bindings decorated with an eight-pointed-star pattern are Morgan M.569, 575, 585, 590, and 663bis2, Ambrosianus Syrus C. 313 inf, Vienna P. Vindb. G. 30501, and Cairo Coptic Museum Hamuli H and Hamuli D. A further elaboration in which each of the eight points of the star produces an extra square is represented by Morgan Library bindings M.574 and 584, British Library Or. 3367, and Berlin Egyptian Museum P. 14023.

43. For the Dura-Europos graffito, see Schmidt-Colinet, "Deux carrés entrelacé," 24.

44. On the development of interlaced patterns before they were used in bookbinding, Trilling notes that "between the second and sixth centuries, innovation took place exclusively in mosaics, especially floor mosaics, and in textiles. After that, architectural sculpture in stone and stucco, opus sectile, metalwork, and manuscript illumination played leading roles in the evolution of Western Medieval, Byzantine, and Islamic interlace" ("Medieval Interlace Ornament," 61). On the two interlaced-square patterns and their history and symbolism, see Schmidt-Colinet, "Deux carrés entrelacé."

45. The Ambrosiana binding must have originally belonged to one of two volumes of the Old and New Testament, as indicated by the title *Book first of the sacred / scriptures of the Old and / New Testament belongs to / Basilios the deacon*, written in Greek on the two boards.

46. Four-leaf interlaced patterns are considered in Reutersward, "The Forgotten Symbols of God," 79–83 and 91–101; also, briefly, in Reutersward, "The Christian Use of the Tetragram." In both cases the author calls the motif a "tetragram" ("for convenience," as he says), which is confusing because the term is well established as meaning something completely different: a word composed of four letters, from the Greek *tetragrammaton*, originally a Hebrew theonym transliterated in Latin as the four-letter word YHWH.

47. Reutersward, "The Forgotten Symbols of God," 92.

48. Ibid., 95. This pattern is found in two bindings from the Morgan Library (M.604 and 608), one from the Coptic museum in Cairo (Cairo Hamuli find I1), and also in Sinai codex Georgian 37 (tenth century) and Sinai codex Georgian 26 (ninth–tenth century).

49. See Maguire "Magic and the Christian Image," 60–61.

50. Ibid., 65.

51. For the use of knots in spells, see Trilling, "Medieval Interlace Ornament," 70.

52. Ibid., 73. See also Maguire, "Magic and Geometry."

53. Trilling, "Medieval Interlace Ornament," 74.

Nine

The Fastenings

No one shall leave his book unfastened when
he goes to the synaxis or to the refectory.

—The Rules of St. Pachomius, fourth century AD

I conographical evidence from a wide range of late antique art indicates
that most codices in late antiquity had some kind of fore-edge flap as
well as various types of fastening (figs. 101, 102, and 103; see also fig.
110; figs. 41 and 42, chap. 4). Fastenings were apparently an essential
appendage, even in single-gathering codices such as the Nag Hammadi
codices. The flaps and fastenings served primarily to protect the book block
and prevent it from remaining open and exposed to the environment. Flaps
might have also been used to further protect and prevent the accumulation
of grime on the fore edge of the leaves, the most heavily handled part of the
book block. The earliest physical evidence we have of fore-edge flaps in multigathering codices is from later Coptic codices, such as in the British Library
binding of Coptic Psalter Or. 5000 (ca. seventh–eighth century AD) and Pap.
1442 (eighth century), followed by many later Armenian and Islamic bound
codices.[1] For fastenings, besides iconographical evidence, we also have phys-

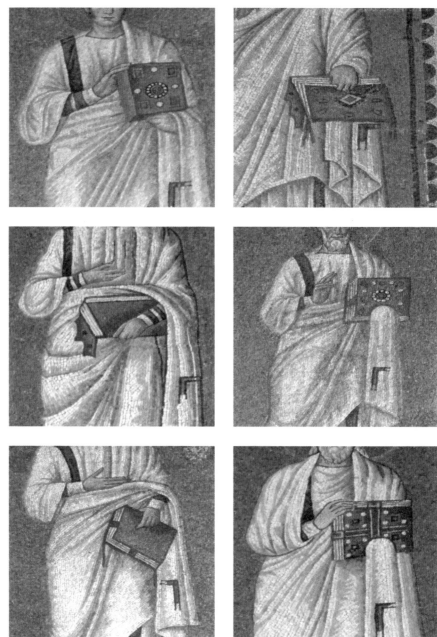

Fig. 101 Prophets with codices. Details from wall mosaics, early 6th century AD, Basilica of Sant'Apollinare Nuovo, Ravenna, Italy.

Fig. 102 St. Laurence with open codex, mosaic, 5th century after AD 425. South arm lunette, Mausoleum of Galla Placidia, Ravenna, Italy. Fore-edge flap and long straps, possibly corner straps, are visible.

Fig. 103 St. Petronilla introducing St. Veneranda to Heaven, 4th century AD. Wall painting, Catacombs of Domitilla, Rome, Italy. Besides an open *capsa* containing several rolls, an open codex with a very long strap is shown in the background.

ical evidence from the earliest complete codices that have been preserved, those in the Chester Beatty Library, the Glazier codex in the Morgan Library and Museum (MS G.67), and the Freer Gospel (Freer Gallery of Art, F1906.297). On Chester Beatty codices 813 (sixth century) and 814 (seventh century), there are straps—still preserved, although extremely fragmented—laced through the fore and head edges of the left board that would have been long enough to wrap around the book two or three times (figs. 104 and 105).

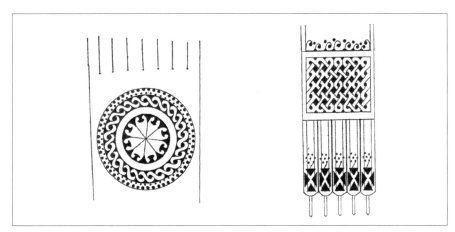

Fig. 104 Fastenings with ink and cutout decorations, detail. Chester Beatty codex 813. From Lamacraft, "Early Book-Bindings from a Coptic Monastery," figs. 3 and 4.

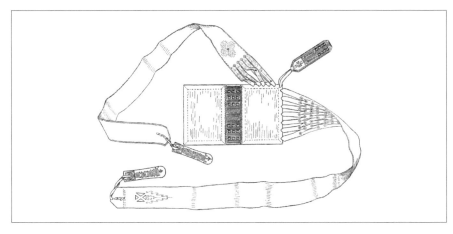

Fig. 105 Reconstruction of Chester Beatty codex 814, showing boards, spine cover, bookmark, and fastening straps. From Petersen, "Early Islamic Bookbindings and Their Coptic Relations," fig. 8.

The fore-edge strap of codex 813 would have measured about 40 inches (1 m) long; the one at the head edge, about 30 inches (80 cm). Similar straps were used in codex 814. In both codices the straps were cut very precisely so that they could expand in a way similar to the slits cut in leather loincloths in ancient Egypt and sandals in Roman Egypt (fig. 106).[2] Once the straps were wrapped around the books, they were secured in place with bone pegs threaded at the end of the straps and decorated with concentric rings (figs. 107, 108, and 109; see also fig. 45, chap. 5) often painted in red and black. The leather straps were also decorated with motifs drawn in black ink, as shown in figures 104 and 105. The two fastening straps wrapped around the codex, one on the vertical axis, the other horizontal, forming a cross on the closed book—a feature that can be extensively observed in representations of books in late antique art (fig. 110; fig. 45, chap. 5). A simpler but similar long strap is still preserved in excellent condition, although detached from the book, with the Glazier codex (Morgan MS G.67). The only evidence that remains of the original straps of the Freer Gospel (Freer Gallery of Art, F1906.297), presumably similar to those preserved in the Chester Beatty codices 813 and 814, are the holes at the fore and head edges through which the straps were originally attached to its left board.

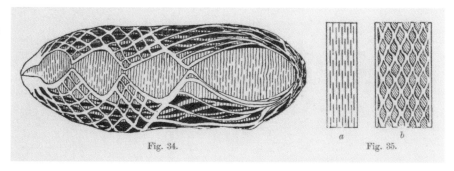

Fig. 106 Late antique sandal from Akhmin-Panopolis, Egypt. Mesh effect produced by slits cut through the leather for expansion. From Frauberger, *Antiken*, figs. 34 and 35. Göttingen State and University Library.

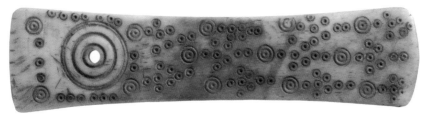

Fig. 107 Peg for a bookbinding fastening strap, ca. 5th–7th century AD. Bone; incised, tinted, and inlaid with paint. The Metropolitan Museum of Art, Rogers Fund, 1907, 07.228.45. Cat. 46.

Fig. 108 Peg for a bookbinding fastening strap, ca. 5th–7th century AD. Bone; incised, tinted, and inlaid with paint. The Metropolitan Museum of Art, Rogers Fund, 1907, 07.228.46. Cat. 47.

Fig. 109 (Left) Peg for a bookbinding fastening strap, ca. 5th–7th century AD. Bone; incised and inlaid with paint. The Metropolitan Museum of Art, Gift of the Egypt Exploration Fund, 1903, 03.4.55. Cat. 48.

Fig. 110 (Right) Icon with effigy of Mark the Evangelist with codex secured with fastening straps, Fayoum, Egypt, 6th century AD. Wood (sycamore), encaustic painting. Paris, Bibliothèque Nationale de France, cabinet des Médailles, FR 1129a.

Later Eastern Mediterranean codices have either interlaced or toggle-and-loop fastening devices. Interlaced fastenings consist of an interlaced strap (bi- or tripartite) with a metal ring at the end (fig. 111) that is laced through one board and a pin in the opposite board.[3] Although interlaced fastenings were typical in Byzantine, Georgian, and Syriac bindings, there are apparently no examples of such rings still preserved on any surviving binding from earlier than the tenth century. The straps were laced through one of the two boards (through the upper board for the Morgan Coptic codices), with their ends usually left untrimmed on the inner face of the boards (see fig. 48, chap. 5). The other component of these fastenings is a pin, usually made of bone or metal, the flattened body of which was slotted inside the fore edge of the opposite board and often secured with a nail running through the thickness of the board and into a hole in the fastening pin (fig. 112; see also fig. 77, chap. 8).[5] The number of fastenings in a single volume generally depended on the size of the book. For large volumes, such as

Fig. 111 (Left) Bipartite fastening strap with metal ring showing interlacing method.

Fig. 112 (Right) Bone fastening pin on Morgan M.577, ca. 9th–10th century AD. Reconstructed metal securing pin.

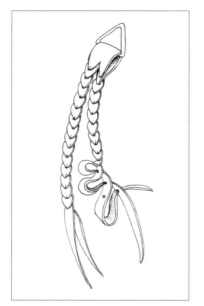

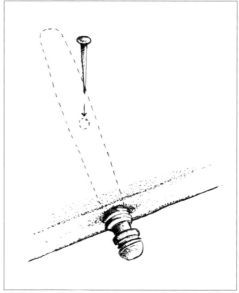

Morgan M.569 and the tenth-century Sinai codex Georgian 63, as many as seven straps could be used, divided between three triple fastening straps on the fore edge and two each for the head and tail edges, while only two or even one on the fore edge would be used for smaller codices. Besides the Morgan Coptic bindings and the tenth-century Sinai Georgian bindings, there are examples of interlaced strap-and-pin fastenings from the Deir el-Bachit monastery in western Thebes that can be dated between the sixth and the ninth or tenth centuries.[6] The same type of fastening was used until well into the seventeenth century in Byzantine and post-Byzantine bookbindings, and the technique itself was widely used for other purposes. From the leather straps with which the Tollund man was hanged in Denmark sometime in the fourth to third century BC (fig. 113) to Roman and late antique sandals (figs. 114 and 115), especially from the second century AD on, and the sixth-century leather aprons from the St. Epiphanius monastery, the technique is identical (see figs. 116, 117, and 118).[7]

Fig. 113 Tollund man, hanged by an interlaced leather strap, ca. 4th–3rd century BC. Silkeborg Museum, Denmark.

Fig. 114 Sandal, left shoe, sole decorated with a stag, 401–600 AD. Egyptian. (war loss). Leather. Vorderasiatisches Museum, Staatliche Museen, Berlin, Germany, 4835.

Fig. 115 Sandal, Egypt, ca. 7th–8th century AD. Leather. Peabody Museum, Yale University, 006676. Cat. 41.

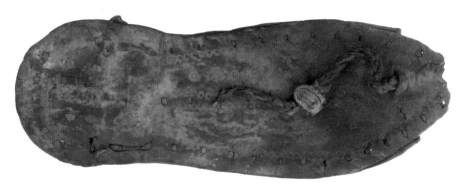

Fig. 116 Fragment of a fastening strap originally on MS M.597, Egypt, 10th century AD. The Morgan Library and Museum, Purchased by J. Pierpont Morgan, 1911, MS M.597. Cat. 40.

Fig. 117 Fragment from a sandal, AD 300–700. Leather. Courtesy Penn Museum, 2003-34-525.1 Cat. 42.

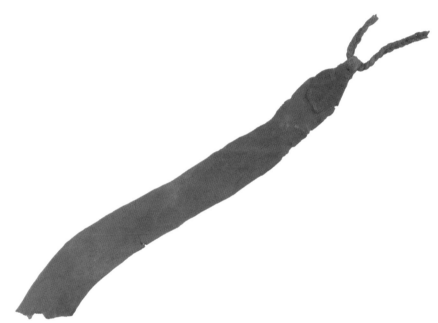

Fig. 118 Fragment of a belt, found at the Monastery of St. Epiphanius, ca. AD 600. Leather. Courtesy of the Division of Anthropology, American Museum of Natural History, 95/2402 D02. Cat. 43.

Toggle-and-loop fastenings are also composed of two elements: (1) a loop made of a leather thong that is laced through two holes in one board and (2) a leather strap with a leather toggle or button at one end, which fits into the loop; the other end of the strap is laced through a hole or slit in the second board and pasted on its inner face. Each book normally had two or three such loops on the fore edge and one at the head and tail edges (fig. 119).[8] Toggles from these fastenings are very rarely preserved; the earliest example we have for a book appears to be on a manuscript discovered in a burial in Naqlun from the end of the eleventh century (fig. 120). Toggle-and-loop fastenings have, however, been used extensively in footwear since antiquity.[9]

Fig. 119 Toggle-and-loop fastening.

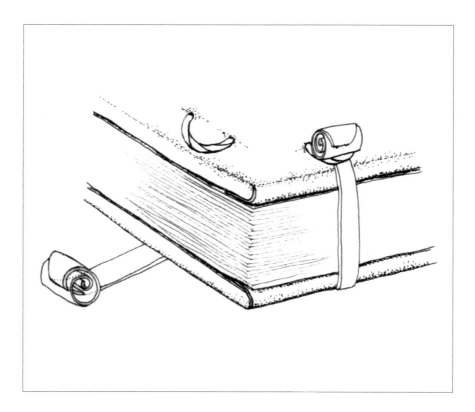

All the fastenings described in this chapter were prone to breaking off the covers, and most of them have been long lost from the books on which they were originally used. In a letter from the sixth or seventh century written on an ostracon now at the Louvre, the writer "requests a monk to bring along his tool to repair the bands of torn books on his next visit."[10]

Fig. 120 Coptic codex, Naqlun, Egypt, AD 1099–1100 (Nd.02.239), from burial T. 324. Archive of the Polish Centre of Mediterranean Archaeology (PCMA).

Notes

Epigraph: Rule 100; translation Armand Veilleux, *Pachomian Koinonia*, 2:161.

1. For a concise treatment of Islamic bindings, see Szirmai, *Archaeology of Medieval Bookbinding*, 51–61. For Armenian bindings, see Merian, "The Structure of Armenian Bookbinding," 61–64.

2. For Chester Beatty codices 813 and 814, see Lamacraft, "Early Book-Bindings from a Coptic Monastery," 224–226, 229–231, figs. 3–7, and plate 2. For the Egyptian leather loincloths, see Veldmeijer, *Sailors, Musicians and Monks*, cat. nos. 71 and 72, and fig. 28; for the use of the same technique in sandals from late antiquity, see Frauberger, *Antike und frühmittelalterliche Fussbekleidungen*, 26, figs. 34 and 35.

3. Such interlaced fastenings are found in the following Morgan Library Coptic bindings: M.569A, 574A, 577A, 580, 587A, 597A, 600A, 604A, and 608A. All the straps are bipartite.

4. In the 1920s, however, the rings were supposedly still preserved on a binding from the eighth or ninth century. See Adam, "Die griechische Einbandkunst," 21, 63, and figures on 78, 79, and 84. A graphic reproduction is also given in Petersen, "Early Islamic Bookbindings," fig. 6; a photograph (not showing the fastenings) appears in Grohmann and Arnold, *Denkmäler islamischer Buchkunst*, 19a.

5. Except for bindings M.569 and 672d, in which the pins are made of copper alloy, all other pins preserved in the Morgan bindings are bone. See also the metal pins preserved on British Library Or. 6801; also shown in Lindsay, "The Edfu Collection of Coptic Books," fig. 21. For twenty-eight examples from the Kairouan bindings, see Marçais and Poinssot, *Objets kairouanais, IXe au XIIIe siècle*, 18, fig. 2.

6. See Veldmeijer, *Sandals, Shoes and Other Leatherwork*, cat. nos. 149, 157, and 171.

7. On the Tollund man, see http://www.tollundman.dk/krop.asp. For sandals, see Veldmeijer, *Sandals, Shoes and Other Leatherwork*, 20–21 (with several other examples mentioned); Frauberger, *Antike und frühmittelalterliche Fussbekleidungen*, cat. nos. 10, 13; Montembault and Musée du Louvre, *Catalogue des chaussures*, cat. nos. 50 (n.d.) and 51 (Coptic period, AD 395–641), 108, 109; see also cat. no. 401 for interlaced straps made with two different colors. For the aprons from the Monastery of St. Epiphanius, see Winlock, *The Monastery of Epiphanius at Thebes*, pt. 1, 76–78. Petersen mentions that two of the Morgan Coptic bindings (M.574A and 577A) have their fastening straps braided of three rather than two strands ("Coptic Bookbindings," 84), but an inspection of the bindings shows that they are consistent with the other bipartite fastenings.

8. Coptic bindings from the Morgan collection with toggle-and-loop fastenings are M.575A, 581A, 583A, 585A, 586A, 588A, 594A, 595A, 596A, 599A, 609A, and 633A.

9. See Frauberger, *Antike und frühmittelalterliche Fussbekleidungen*, 21, 22, and fig. 25. See also Szirmai, *Archaeology of Medieval Bookbinding*, 41, fig. 3.9; Veldmeijer, *Sandals, Shoes and Other Leatherwork*, 161, cat. no. 178, and *Sailors, Musicians and Monks*, 64, fig. 34, and cat. no. 55.

10. Kotsifou, "Bookbinding and Manuscript Illumination," 227.

The Bookmarks and Board Corner Straps

Closing the book, then, and putting my finger or something else for a mark I began—now with a tranquil countenance—to tell it all to Alypius.

—St. Augustine, *Confessions*, AD 397–400

R eading a text is a dynamic process in which readers often experience interruptions, after which they must retrieve a specific passage. Bookmarks are a great help in this process because they can mark the page where the reading was interrupted or a standard stopping place, such as the beginning of a chapter, as well as a point of major interest. They would have been impractical with rolls or wax tablets, and there is no evidence that they were used until the advent of codices. In fact, the easy retrieval of specific passages has been proposed as one of the practical advantages that may have contributed to the gradual establishment of the codex as a book format.[1] There are three types of bookmark: movable (the marker has one part permanently connected to the book and another part that the reader can place anywhere within it), fixed (the marker is permanently connected to a particular leaf and cannot be moved), and free (the marker is a loose piece of material without any permanent connection to the book). Board strap markers and endband string markers are movable; leaf tab markers and leaf string markers are fixed. Because of their nature, free markers are rarely found in early codices, and even if found, they are usually impossible to date with accuracy.

Leaf markers were used for permanently marking a specific leaf of a codex. They consisted of either textile or leather tabs (leaf tab markers) pasted at the fore edge of the specific leaf to be marked (fig. 121) or thread (leaf thread markers) sewn through the leaf and twisted so that the twisted part extends from the fore edge of the book block (fig. 122). Leaf tab markers can be found as early as Chester Beatty codices 813, 814, and 815 and were widely used in codices around the Mediterranean until as late as the seventeenth century.[2] Leaf thread markers have been documented in Syriac, Ethiopic, Greek, and Armenian manuscripts as early as the fourth century, with Codex Sinaiticus representing one of the earliest examples, although it is not certain when the markers were sewn through the leaves of the codex.

Board strap markers consist of leather straps that are laced—or often nailed or pasted—in the inner face of the fore edge of the book boards so that their extending part can be inserted between the leaves of a closed book block, marking temporarily a specific opening or leaf (fig. 123; see also fig. 5, introduction). These straps have been found in most of the Eastern

Fig. 121 (Left) Leather leaf tab marker (right edge). Sinai codex Syriac 30, fol. 91v, AD 778, probably rebound at later date. St. Catherine's Monastery, Sinai, Egypt.

Fig. 122 (Right) Leaf thread markers. Fore edge, Sinai codex Syriac 75, AD 1295. St. Catherine's Monastery, Sinai, Egypt.

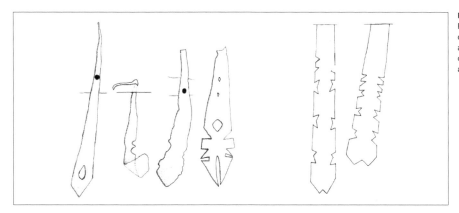

Mediterranean bookbinding traditions, but in all but the Sinai bindings they are broken off, with only the nailed or pasted edges surviving on the inner face of the boards. Although Petersen erroneously identified the markers in Morgan Coptic bindings as "lifting tabs" (that is, tabs for lifting the boards),[3] they can be identified as marking devices not only from iconographical evidence but also from stains left between the leaves of more or less intact codices in the St. Catherine's Monastery library and other collections. Although their use peaked in the thirteenth century, board strap markers were still commonly used until at least the fourteenth or fifteenth century.[4] Based on available evidence, endband string markers—that is, threads, cords, or ribbons sewn or otherwise fastened to headbands—developed much later than the late antique codices we have been considering.

Besides the board strap markers in some of the earliest preserved codices, for example, in Chester Beatty codices 813 and 814, there are straps extending from the corner of one of the two boards that look as if they might have been bookmarks. They consist of tags made of leather and parchment, decorated with indented edges, blind tooling, cutout patterns, and inked designs (fig. 124; see also fig. 105, chap. 9).[5] Starting with the earliest surviving codices, such as the Glazier codex (Morgan MS G.67) and the Freer Gospel, and extending to the ninth- and tenth-century Coptic bindings of the Morgan collection, as well as Georgian, Greek, and Syriac bound codices of the St. Catherine's Monastery, similar holes are found in the upper outer corners of both the left and right boards, not just in only one of the two. In several of the Morgan Coptic bindings, the holes in each board are double and still preserve the ends of interlaced straps inserted through them (see fig. 5, introduction; fig. 46, chap. 5).[6] Because there is no example preserving anything but these strap ends, we cannot clearly identify their purpose. The board corner straps would probably not have been bookmarks, however, because many of these codices already had leaf tab or string markers or else board strap markers.[7] The corner straps are often double, thicker and stronger than the straps that served as bookmarkers, which suggests that their purpose demanded extra strength. From iconographical evidence, it appears that these straps could be connected together, as shown in a seventh-century icon from Sinai (fig. 125, see also fig. 102), possibly so that the books could

Fig. 124 (Left) Bookmark, recto and verso, originally on CPT 814, 6th century AD. © The Trustees of the Chester Beatty Library, Dublin, CPT 814.

Fig. 125 (Right) Christ enthroned, 7th century AD. St. Catherine's Monastery, Sinai, Egypt. Red straps connecting upper corners of open codex are visible. Note also tripartite straps with rings (right board) and corresponding pins (left board).

be hung.[8] An illustration from Robert Curzon's 1865 *Visits to the Monasteries of the Levant* shows books that are placed not on the shelves overhead but rather inside leather bags, similar to those used for Ethiopic codices, that are suspended from wooden poles protruding from the wall (figs. 126, 127). Perhaps the corner straps were a similar device for hanging a book; in any case, they appear to have been completely abandoned after the tenth century.

Fig. 126 Interior of Abyssinian Library, Monastery of Souriani on the Natron Lakes. From Robert Curzon, *Visits to Monasteries of the Levant*, 137.

Fig. 127 Codex of Psalter and other texts with satchel, Ethiopia, 18th century AD. The Morgan Library and Museum, Gift of David McC. McKell, 1961, MS M.911. Cat. 18.

Notes

Epigraph: Book 8, chap. 12, 30; translation from https://www.ourladyswarriors.org /saints/augcon8.htm.

1. For convenience of use, see Gamble, *Books and Readers*, 55–56.
2. See Lamacraft, "Early Book-Bindings," 217. For the leaf tab markers among the Morgan Coptic codices, see Petersen, "Coptic Book-bindings," 94.
3. See Petersen, "Coptic Bookbindings," 92–93. Board strap markers are found in the following Morgan Coptic bindings: M.569, 574, 577, 580, 586, 587, 588, 594, 595, 597, 599, 600, 604, 608, 609, 634.
4. For a discussion of board strap markers, see Boudalis, "Clarifying the Structure," 288–294.
5. For the bookmarks in Chester Beatty codices, see Lamacraft, "Early Book-Bindings from a Coptic Monastery," 226, 231, plates 4c and 5b,c. For another example, see the reproduction in Wulff, *Altchristliche und mittelalterliche byzantinische und italienische Bildwerke*, 155, item 68, plate 30.
6. Double holes in each board corner among the Morgan Coptic bindings are found in M.587, 590, 597, 600, 601, and 608, while in M.633 the holes are single on each board. In M.574 there was apparently a cord rather than a leather strap laced through the corner holes. To my knowledge, no examples with double holes on each board have been documented in the Sinai.
7. For further discussion of board corner straps, see Boudalis, "Clarifying the Structure," 295–299, where these corner straps are called "board head edge straps."
8. For further iconographical evidence, see Boudalis, ibid., 269, fig. 162.

Conclusion

Between the second and sixth centuries AD, the format for written texts changed from the roll to the multigathering codex—or the book as we know it today. The aim of *The Codex and Crafts in Late Antiquity*, the parallel exhibition, and the research on which they are based has been to identify not why this change happened but rather how it did by casting light on the antecedents and possible provenance of the various techniques used to make the multigathering codex and its constituent parts. To answer this question, the research evolved in two parallel directions: comparing the multigathering codex with simpler types of codices that predate it (wooden tablet and single-gathering codices) and comparing the multigathering codex to other artifacts of the same historical and cultural context (late antiquity).

We have found that the multigathering codex does combine features borrowed from both the wooden tablet codex and the single-gathering codex. Sewing through the fold is probably inherited from the single-gathering codex, while sewing multiple gatherings into a single structure appears to have been adapted from the wooden tablet codex. Sewing holes arranged in pairs and sewing these pairs independently is a feature common to all three types of codex. In the multigathering codex, however, the use of paired sewing holes, with each pair sewn independently, appears to derive from the wooden tablet codex, in which multiple units (in this case tablets) also make up the book block.

The comparison of techniques used for making the multigathering codex with those used in other crafts of late antiquity, supplemented by documentary and iconographical evidence, serves to place the construction of the multigathering codex within the technological background of late antiquity. The evidence points to a close relation between the techniques employed in making the multigathering codex and those used in a variety of everyday crafts and common objects, such as woven textiles, baskets, mats, socks, shoes, and sandals. Most if not all of the components of the multigathering codex, both structural and decorative, can be directly or indirectly related to a wide range of crafts and artifacts: the sewing employed the same technique used to make socks and possibly more elaborate fabrics; the endbands were fashioned with techniques used for the edge finishing of textiles; and the decoration of the covers and the fastening straps borrowed techniques commonly used in shoes, sandals, belts, and other leatherwork.

There are many references in early Christian literary sources and in accounts of archaeological finds from cells, hermitages, and monasteries in Egypt indicating that basketry, mat making, weaving, and leatherworking were all part of the repertoire of technical skills and crafts of early Christian monks, who are considered responsible for producing most of the surviving early Christian books.[1] In his *Lausiac History*, written in AD 419–420, Palladius describes the activities of the monks of Nitria: "One works the land as a farmer, another the garden, another works at the forge, another at the bakery, another in the carpenter's shop, another in the fuller's, another weaving the big baskets, another in the tannery, another in the shoe-shop, another at calligraphy, another weaving the soft baskets. And they learn all the Scriptures by heart."[2]

Although asceticism was rather common among various groups of both pagans and Christians in Egypt during the first three centuries AD, by around the end of the third century and the first decades of the fourth century monasticism as we know it started to take shape in Egypt and spread very rapidly. The first monastery in Constantinople was founded in 382, and by 536 there were about seventy monasteries in the capital of the Byzantine Empire alone.[3] In Egypt, there were major monastic centers in the Nitrian Desert, about thirty miles south of Alexandria; in Kellia, further into the desert; and also in Akhmin-Panopolis and Scetis.[4] Ever since its foundation, monasticism has been a central feature of Christianity, in both the East and the West.

Monks inside monastic communities undertook various occupations besides copying books—and presumably binding them. Activities such as those Palladius mentions are well documented for Egyptian monasteries through archaeological research as well as in most of the early literature on the desert fathers, including not only the *Lausiac History* but also the *Alphabetic Anthology* compiled in the fifth and sixth centuries. The artifacts the monks produced were sold outside the monasteries to provide income for the community, while in other cases, such as producing manuscripts, monks worked on commission. According to Ewa Wipszycka, the monks' main occupations were primarily basketry, including mats, nets, and ropes, followed by weaving and manuscript copying and binding. She writes that "in the light of both literary and documentary texts there is no doubt that many monks performed several crafts at a time."[5] Pesynthios, later to become bishop of Hermonthis (present-day Armant), who learned to copy and bind manuscripts when he was eleven, was also a mason and a carpenter.[6] Frange, a monk living in western Thebes during the late seventh to early eighth century, was weaving shrouds and thin bands called *keiriai* (used as bandages for wrapping around shrouded corpses), manufacturing ropes, and also copying and binding books on a "professional" basis, as his extensive archive of ostraca indicates.[7] Many such ostraca written in Coptic contain information about the practice of writing and binding books.[8] From the available documentary evidence, it appears that the cost for binding a manuscript could equal the cost of copying it, which may be one of the reasons that it is not rare to find reused binding boards and covers.[9]

It should not be surprising that individuals with multiple skills as well as craftsmen working in close proximity would transfer techniques from one craft to another. As with other crafts, such as the diffusion of window glass, organized religion was a major force behind the development of the codex.[10] Luke Lavan has noted that some scholars of medieval history "have considered monasteries to be equivalent to Roman forts as technological showcases, perhaps being responsible for the spread of water-mills to Ireland and other innovations."[11] Although the great majority of all the surviving codices from late antiquity must have been written and bound in a monastic milieu, there were certainly books produced in secular settings too, especially in earlier centuries. It would be interesting to know whether there were differences in the ways religious and secular books were bound and decorated. This, however, is one of the many questions that for the moment will have to remain unanswered.

Notes

1. On book production in early monastic communities, see, for example, Kotsifou, "Books and Book Production," 48–66; Schachner, "Book Production in Early Byzantine Monasteries," 21–30. For literary and archeological evidence on crafts in monastic communities, see Wipszycka, "Les aspects économiques de la vie de la communauté des Kellia" and "Resources and Economic Activities of the Egyptian Monastic Communities." See also Winlock and Crum, *The Monastery of Epiphanius at Thebes*, pt. 1; Boud'hors and Heurtel, "The Coptic Ostraca from the Tomb of Amenemope."

2. Quotation from Veilleux, *Pachomian koinonia*, 2:129.

3. See Talbot, "Monasticism," 1392–1394.

4. On Egyptian monastic centers, see Timbie's "Egypt" entry in *Encyclopedia of Monasticism*.

5. Wipszycka, "Resources and Economic Activities," 183.

6. Ibid.

7. On Frange, see Heutrel, "Que fait Frange?"; Boud'hors, "Copie et circulation des livres." On monks' weaving practices, see Wipszycka, "Resources and Economic Activities," 173–182.

8. For a discussion of evidence from ostraca, see Kotsifou, "Books and Book Production," and "Bookbinding and Manuscript Illumination."

9. On the pricing of manuscripts usually not including the cost of the binding, see Boud'hors and Heurtel, "Copie et circulation des livres," 160; see also Bagnall, *Early Christian Books in Egypt*, 57–58, table 3.1.

10. Lavan, "Explaining Technological Change," xxvii.

11. Ibid.

Checklist of Objects
in the Exhibition

Cat. 1 (see fig. 8)
Tablet codex, 4th century AD
Exercise tablets belonging to a schoolboy
Egypt
Wood, wax
Various dimensions, average: 6¾ × 5 3⁄16 × ¼ in.
(17.2 × 13.2 × 0.7 cm)
Brooklyn Museum, CUR.37.1908E, 37.1909E,
37.1910E, 37.474E, 37.473E

Cat. 2
Facsimile of a tablet codex, 2016
Based on 4th-century AD exercise tablets
(see cat. 1)
Georgios Boudalis, maker
Cedar wood, linen string, colored wax
Closed: 7 × 5½ × 1¼ in. (17.8 × 14 × 3.2 cm),
open: 7 × 11 × ¾ in. (17.8 × 27.9 × 1.9 cm)
Courtesy of Georgios Boudalis

Cat. 3 (see also fig. 11)
Facsimile of a tablet codex, 2016
Based on a tablet codex found in the House of
the Bicentenary in Herculaneum
Georgios Boudalis, maker
Cedar wood, linen string, colored wax
Closed: 5¼ × 2¾ × 1¼ in. (13.3 × 7 × 3.2 cm),
open: 5¼ × 5½ × ½ in. (13.3 × 14 × 1.3 cm)
Courtesy of Georgios Boudalis

Cat. 4 (see also fig. 10)
Facsimile of a tablet codex, 2016
Based on a drawing published in Wilhelm
Schubart, *Das Buch bei den Griechen und
Römern*, 16, fig. 1.
Georgios Boudalis, maker
Cedar wood, linen string, colored wax
3⅓ × 1 9⁄16 in. (13.5 × 4 cm)
Courtesy of Georgios Boudalis

Cat. 5 (see also fig. 9)
Facsimile of a tablet codex, 2016
Based on eight tablets of a ca. AD 360 "Farm
Account Book" found in the Dakhleh Oasis, Egypt
Georgios Boudalis, maker
Cedar wood, hemp string
13 × 4¼ × 1¼ in. (33 × 10.75 × 3.18 cm)
Courtesy of Georgios Boudalis

Cat. 6 (see fig. 7)
Kylix (drinking cup), ca. 460 BC
Attributed to the painter of Munich 2660
Greece, Attica
Terracotta (earthenware), red-figure technique
2¾ × 7⅞ × 7⅞ in. (7 × 20 × 20 cm)
Metropolitan Museum of Art, Greek and Roman,
Rogers Fund, 1917, 17.230.10

Cat. 7
Single-gathering codex inner structure, 2017
Based on the codices found in Nag Hammadi,
Egypt, dated second half of the 4th century AD
Georgios Boudalis, maker
Paper, leather, linen string
9⅞ × 7 × ¾ in. (25 × 18 × 2 cm)
Courtesy of Georgios Boudalis

Cat. 8
Facsimile of single-gathering codex, 2017
Based on the codices found in Nag Hammadi,
Egypt, dated second half of the 4th century AD
Georgios Boudalis, maker
Tanned leather (goat), paper, linen string
Each of two, closed: 9⅞ × 7⅞ × ¾ in. (25 × 20 ×
2 cm), open: 9⅞ × 19¾ in. (25 × 50 cm)
Courtesy of Georgios Boudalis

Cat. 9
Single-gathering codex sewn with stab sewing,
2017
Georgios Boudalis, maker
Paper, linen string
9⅞ × 7 × ⁹⁄₁₆ in. (25 × 18 × 1.5 cm)
Courtesy of Georgios Boudalis

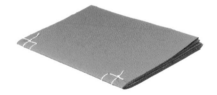

Cat. 10
Single-gathering codex sewn with overcasting,
2017
Georgios Boudalis, maker
Paper, linen string
9 ⅞ × 7 × ⅜ in. (25 × 18 × 1 cm)
Courtesy of Georgios Boudalis

Cat. 11 (see fig. 33)
Textile fragments made with cross-knit looping,
possibly from a sock, 323 BC–AD 256
Dura-Europos, Syria
Wool; dyed three different colors
3 ¼ × 1 ⅜ in. (8.2 × 3.5 cm), 2 ⅜ × 2 ⅜ in. (6.2 × 6.2 cm)
Yale University Art Gallery, Yale-French
Excavations at Dura-Europos, 1935.556

Cat. 12 (see fig. 32)
Textile fragment made with cross-knit looping,
ca. AD 200–256
Dura-Europos, Syria
Wool
5 ⅞ × 6 ¹¹⁄₁₆ in. (15 × 17 cm)
Yale University Art Gallery, Yale-French
Excavations at Dura-Europos, 1933.483

Cat. 13 (see fig. 34)
Replica of a sock and sock starter, 2016
Made with the cross-knit looping technique after
a 4th–5th century AD sock from Fayum, Egypt
Regina de Giovanni, maker
Acrylic yarn; dyed three different colors
Sock: 4 × 9 × 3 ½ in. (10.2 × 22.9 × 8.9 cm);
starter: ¾ × 4 × 2 (1.9 × 10.16 × 5.1 cm)
Courtesy of Regina de Giovanni

Cat. 14
Sample of cross-knit looping technique, 2017
Georgios Boudalis, maker
Hemp cord
8 ⅝ × 6 ½ in. (22 × 16.5 cm)
Courtesy of Georgios Boudalis

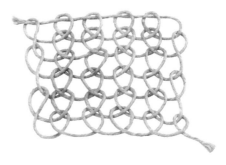

Cat. 15
Multi-gathering codex sewn with stab sewing, 2017
Georgios Boudalis, maker
Paper, linen string
9 ⅞ × 7 × ¾ in. (25 × 18 × 2 cm)
Courtesy of Georgios Boudalis

Cat. 16
Book block cut to reveal loop stitching inside the
centerfold, 2017
Georgios Boudalis, maker
Paper, linen string
9 ⅞ × 7 × ⅞ in. (25 × 18 × 2.2 cm)
Courtesy of Georgios Boudalis

Cat. 17 (see fig. 96)
Bookbinding fragment, lower cover,
9th–10th century AD
Originally on MS M.574, Hermeneiai with Various
Hymns
Fayum, Egypt
Leather over papyrus board; parchment
11⅛ × 10 in. (28.3 × 25.5 cm)
Morgan Library and Museum, Purchased for
J. Pierpont Morgan, 1911, MS M.574A2

Cat. 20
Book blocks, showing different thread-hinging
techniques, 2016
Georgios Boudalis, maker
Various woods, paper, linen string
Each of three, closed: 6 × 4¼ × ½ in. (15.2 × 10.8 ×
1.2 cm)
Courtesy of Georgios Boudalis

Cat. 18 (see fig. 127)
Codex of Psalter and other texts with satchel,
18th century
Ethiopia
Vellum, wood, string; leather
7 × 5 in. (18 × 12.7 cm)
Morgan Library and Museum, Gift of David McC.
McKell, 1961, MS M.911

Cat. 19 (see fig. 45)
Facsimile of the Glazier Codex, 2011
Based on the original from the 5th century AD,
now in the Morgan Library and Museum (MS G. 67)
Ursula Mitra, maker
Oak wood, leather (goat), paper, parchment,
linen thread, bone
Closed: 5 × 4¼ × 1⅞ in. (12.7 × 10.8 × 4.7 cm),
open: 5 × 9¼ × 1⅜ in. (12.7 × 23.5 × 3.5 cm)
Courtesy of Ursula Mitra

Cat. 21 (see fig. 86)
Bookbinding fragment, upper cover,
9th–10th century AD
Originally on MS M.577, Miscellany
Fayum, Egypt
Leather over papyrus board; parchment
13 × 10 in. (33 × 25.5 cm)
Morgan Library and Museum, Purchased for
J. Pierpont Morgan, 1911, MS M.577A1

Cat. 22
Bookbinding fragment, lower cover,
9th–10th century AD
Originally on MS M.599, Homily on the
Holy Cross
Fayum, Egypt
Leather over papyrus board; parchment
13¾ × 12 in. (35 × 30.5 cm)
Morgan Library and Museum, Purchased for
J. Pierpont Morgan, 1911, MS M.599A2

Cat. 23 (see fig. 53)
Tunic front with marine motifs, 6th century AD
Egypt
Wool
13 × 44½ in. (33 × 113 cm)
Brooklyn Museum, Charles Edwin Wilbour Fund,
38.753

Cat. 24 (see fig. 65)
Textile fragment, 5th–6th century AD
Egypt
Wool; tabby weave
3 × 6 in. (7.6 × 15.2 cm)
Brooklyn Museum, Gift of the Egypt Exploration
Fund, 15.475F

Cat. 25 (see fig. 58)
Tunic fragment (band with cross and short-
legged aquatic birds), 5th–7th century AD
Egypt
Wool, linen; plain and tapestry weave; applied
braid
12 × 4½ in. (30.5 × 11.2 cm)
Metropolitan Museum of Art, Islamic Art,
Purchase by subscription, 1889, 89.18.308

Cat. 26
Multi-gathering codex shown without the cover
and fastenings, 2017
Georgios Boudalis, maker
Pine wood, paper, linen thread
Open: 7¼ × 10½ in. (18.5 × 27 cm),
closed: 7¼ × 5⅛ in. (18.5 × 13 cm)
Courtesy of Georgios Boudalis

Cat. 27 (see fig. 84)
Bookbinding fragment, upper cover,
9th–10th century AD
Originally on manuscript MS M.584, Hagiography
Fayum, Egypt
Leather over papyrus board; parchment
6⅜ × 8¾ in. (16.4 × 22.2 cm)
Morgan Library and Museum, Purchased for
J. Pierpont Morgan in 1911, MS M.584A

Cat. 28 (see fig. 98)
Bookbinding fragments, lower cover,
9th–10th century AD
Originally on MS M.604, Homily on Gilead
Fayum, Egypt
Leather over papyrus board; parchment
13⅝ × 10⅜ in. (34.5 × 26.5 cm)
Morgan Library and Museum, Purchased for
J. Pierpont Morgan, 1911, MS M.604A2

Cat. 29 (see fig. 82)
Facsimile of a Coptic binding cover (Petersen
no. 71), 1940s
Based on an original 8th–9th century codex in
the Austrian National Library, Vienna (P. Vindob.
G 30501)
Theodore C. Petersen, maker
Colored paper layers mounted on paper
Pattern: 12½ × 9¼ in. (31.8 × 23.5 cm);
sheet: 15⅛ × 11⁹⁄₁₆ in. (38.3 × 29.4 cm)
Morgan Library and Museum, PCC 93

Cat. 30 (see fig. 95)
Facsimile of a Coptic binding cover (Petersen no. 77), 1940s
Based on an original 8th–9th century AD codex in the Egyptian Museum, Berlin (P. 14018)
Theodore C. Petersen, maker
Colored paper layers mounted on paper
Pattern: 13 3/8 × 7 1/4 in. (34 × 18.4 cm); sheet: 13 3/8 × 10 11/16 in. (34 × 27.2 cm)
Morgan Library and Museum, PCC 94

Cat. 31 (see fig. 93)
Roundel from a tunic fragment, 3rd–4th century AD
Egypt
Wool, linen; plain and tapestry weave
12 5/8 × 13 3/8 in. (32 × 34 cm)
Metropolitan Museum of Art, Islamic Art, Purchase by subscription, 1889, 89.18.85

Cat. 32 (see fig. 94)
Tunic fragment with roundel, 3rd–4th century AD
Egypt
Wool, linen; plain weave, tapestry weave
17 1/4 × 15 3/4 in. (43.8 × 40 cm)
Metropolitan Museum of Art, Islamic Art, Rogers Fund, 1909, 09.50.1459

Cat. 33 (see fig. 78 and below)
Bookbinding fragments, upper and lower covers, 9th–10th century AD
Originally on MS M.569, Gospels
Leather over papyrus board; gilded parchment, thread
15 × 11 1/2 in. (38.2 × 29.5 cm)
Morgan Library and Museum, Purchased for J. Pierpont Morgan, 1911, MS M.569A1,A2

Cat. 34
Shoe, 4th–7th century AD
Akhmim-Panopolis, Egypt
Leather, decorated with gilding
9 1/4 in. (23.5 cm)
Metropolitan Museum of Art, Islamic Art, Gift of George F. Baker, 1890, 90.5.34b

Cat. 35 (see fig. 90)
Pair of shoes, 3rd–9th century AD
Akhmim-Panopolis, Egypt
Leather, decorated with cutout openwork, stitching, gilding, tooling, and lacing; thread
9 1/16 in. long (23 cm)
Metropolitan Museum of Art, Islamic Art, Gift of Edward S. Harkness, 1926, 26.9.11a,b

Cat. 36 (see fig. 87)
Slipper, 3rd–7th century AD
Akhmim-Panopolis, Egypt
Leather
7 in. (17.8 cm)
Metropolitan Museum of Art, Islamic Art, Gift of George F. Baker, 1890, 90.5.37a

Cat. 37 (see fig. 81)
Coiffure support, 3rd–7th century AD
Attributed to Egypt
Leather, decorated with cutout openwork and stitching; thread, textile
9 1/2 × 1 3/4 in. (24.1 × 4.4 cm)
Metropolitan Museum of Art, Islamic Art, Gift of George F. Baker, 1890, 90.5.40

Cat. 38 (see fig. 80)
Shoe appendages, AD 300–700
Egypt
Leather, decorated with cutout openwork and stitching; thread
Left to right: 2 in., 2 in., 1 5/8 in. (5.3 cm, 5 cm, 4.1 cm)
Penn Museum, Philadelphia, Gift of the Civic Center Museum, 2003, 2003-34-315 A–C

Cat. 39 (see fig. 80)
Leather fragment, shoe appendage, AD 300–700
Egypt
Leather, decorated with cutout openwork and
stitching; thread
2 in. (5 cm)
Penn Museum, Philadelphia, Gift of the Civic
Center Museum, 2003, 2003-34-351L

Cat. 40 (see fig. 116)
Fastening strap from a book binding,
10th century AD
Originally on MS M.597, Homilies on the
Incarnation and Virgin Mary
Fayum, Egypt
Leather
2⅛ × ¼ in. (5.5 × 0.6 cm)
Morgan Library and Museum, Purchased for
J. Pierpont Morgan, 1911, MS M.597A2F

Cat. 41 (see fig. 115)
Sandal, 7th–8th century AD
Egypt
Leather
Approx. 4 in. (10 cm)
Peabody Museum, Yale, YPM ANT 006676

Cat. 42 (see fig. 117)
Fragment from a sandal, AD 300–700
Egypt
Leather
9⅞ × 3½ × ¾ in. (25 × 9 × 2 cm)
Penn Museum, Philadelphia, Gift of the Civic
Center Museum, 2003, 2003-34-525.1

Cat. 43 (see fig. 118)
Fragment of a belt, ca. AD 600
Found at the Monastery of Epiphanius, Thebes
Leather
⅜ × 13⅝ × 1⅜ in. (1 × 34.5 × 3.5 cm)
American Museum of Natural History, 95/2402
D02

Cat. 44
Replica of a fastening strap, 2017
Georgios Boudalis, maker
Leather, silver alloy
9 × ¾ in. (23 × 2 cm)
Courtesy of Georgios Boudalis

Cat. 45
Facsimile of a Byzantine book binding, 2007
Based on an original binding from the 14th
century
Ursula Mitra, maker
Cedar wood, leather (goat), paper, linen thread,
brass; blind tooling
9⅛ × 5¾ × 2¼ in. (23 × 14.5 × 5.7 cm)
Courtesy of Ursula Mitra

Cat. 46 (see fig. 107)
Peg for a bookbinding, 5th–7th century AD
Attributed to Egypt
Bone; incised, tinted, and inlaid with paint
4½ × 1⅛ in. (11.5 × 2.8 cm)
Metropolitan Museum of Art, Islamic Art,
Rogers Fund, 1907, 07.228.45

Cat. 47 (see fig. 108)
Peg for a bookbinding, 5th–7th century AD
Attributed to Egypt
Bone; incised, tinted, and inlaid with paint
4¾ × 1⁵⁄₁₆ in. (12 × 3.3 cm)
Metropolitan Museum of Art, Islamic Art,
Rogers Fund, 1907, 07.228.46

Cat. 48 (see fig. 109)
Peg for a bookbinding, 5th–7th century AD
Attributed to Egypt
Bone; incised and inlaid with paint
2⁷⁄₁₆ × ½ in. (6.2 × 1.5 cm)
Metropolitan Museum of Art, Islamic Art, Gift of
Egypt Exploration Fund, 1903, 03.4.55

Bibliography

AA.VV. *Antinoe cent'anni dopo*. Exhibition catalogue. Florence: Istituto Papirologico Vitelli, 1998.

Åberg, Nils. *The Occident and the Orient in the Art of the Seventh Century*. Stockholm: Wahlström and Widstrand, 1943–1947.

Adam, Paul. "Die griechische Einbandkunst und des frühchristliche Buch." *Archiv für Buchbinderei* 24 (1924): 21–24.

Ammirati, Serena. "The Use of Wooden Tablets in the Ancient Graeco-Roman World and the Birth of the Book in Codex Form: Some Remarks." *Scripta* 6 (2013): 9–15.

Arnold, Thomas W., and Adolf Grohmann. *The Islamic Book: A Contribution to Its Art and History from the VII–XVIII Century*. London: Pegasus, 1929.

L'art copte en Égypte: 2000 ans de christianisme. Paris: Institute du Monde Arabe / Gallimard, 2000.

Ashall Glaister, Geoffrey. *Glossary of the Book: Terms Used in Paper-Making, Printing, Bookbinding and Publishing with Notes on Illuminated Manuscripts, Bibliophiles, Private Presses and Printing Societies*. London: Allen and Unwin, 1960.

Atsalos, Basile. *La terminologie du livre-manuscrit à l'époque byzantine*. Pt. 1: *Termes désignant le livre-manuscrit et l'écriture*. Thessaloniki: Hetaireia Makedonikon Spoudon, 1971. Reprint, Thessaloniki: University Studio Press, 2001.

Bagnall, Roger S. *Early Christian Books in Egypt*. Princeton, N.J.: Princeton University Press, 2009.

——. *Egypt in Late Antiquity*. Princeton, N.J.: Princeton University Press, 1993.

Baiardi, Ottavio Antonio, and Pasquale Carcani. *Delle antichità di Ercolano*. 10 vols. Naples: Regia Stamperia, 1755–1831.

Barber, E. J. W. *The Mummies of Ürümchi*. New York: Norton, 1999.

——. *Prehistoric Textiles: The Development of Cloth in the Neolithic and Bronze Ages with Special Reference to the Aegean*. Princeton, N.J.: Princeton University Press, 1991.

Bausi, Alexandro, and Eugenia Sokolinski, eds. *Comparative Oriental Manuscript Studies: An Introduction*. Hamburg: Tradition, 2015.

Bel, Alfred, and Prosper Ricard. *Le travail de la laine à Tlemcen*. Algiers: Jourdan, 1913.

Bender Jørgensen, L. "Stone-Age Textiles in Northern Europe." In *Textiles in Northern Archaeology: NESAT III*, 1–10. London: Archetype, 1990.

Bergman, Ingrid, Hans-Åke Nordström, and Torgny Säve-Söderbergh. *Late Nubian Textiles*. Stockholm: Esselte Studium, 1975.

Bird, Junius Bouton, and Louisa Bellinger. *Paracas Fabrics and Nazca Needlework, 3rd Century B.C.–3rd Century A.D.* Textile Museum, catalogue raisonné. Washington, D.C.: National Gallery of Art, 1954.

Blanchard, Alain, ed. *Les débuts du codex*. Bibliologia 9. Turnhout, Belgium: Brepols, 1989.

Bosch, Gulnar K., John Carswell, and Guy Petherbridge. *Islamic Bindings and Bookmaking*. Chicago: Oriental Institute, University of Chicago, 1981.

Boucher-Colozier, Ét. "Un bronze d'époque alexandrine: Réalisme et caricature." *Monuments et mémoires de la Fondation Eugène Piot* 54, no. 1 (1965): 25–38.

Boudalis, Georgios. "The Bindings of the Early Christian Codex: Clarifying the Coptic Contribution to Bookbinding Structures." In *Bookbindings: Theoretical Approaches and Practical Solutions*, edited by N. Golob and

J. Vodopivec Tomažič, 67–82. Bibliologia 45. Turnhout, Belgium: Brepols, 2017.

——. "Clarifying the Struture, Appearance and Use of the Early Codex around the Mediterranean Basin: The Use of Iconographical Evidence." In *Care and Conservation of Manuscripts 15: Proceedings of the Fifteenth International Seminar Held at the University of Copenhagen 2nd–4th April 2014*, edited by M. J. Driscoll, 287–303. Copenhagen: Museum Tusculanum Press, University of Copenhagen, 2016.

——. "The Conservation of an Early 16th-Century Bound Greek Manuscript: An Insight into Byzantine Bookbinding through Conservation." In *Care and Conservation of Manuscripts 13: Proceedings of the Thirteenth International Seminar Held at the University of Copenhagen 13th–15th April 2011*, edited by M. J. Driscol, 199–214. Copenhagen: Museum Tusculanum Press, 2012.

——. "Endbands in Greek-Style Bindings." *The Paper Conservator* 31 (2007): 29–49.

——. "The Evolution of a Craft: Post-Byzantine Bookbinding between the Late Fifteenth and the Early Eighteenth Century from the Libraries of the Iviron Monastery in Mount Athos/Greece and the St. Catherine's Monastery in Sinai/Egypt." PhD thesis, University of the Arts, London, 2004.

——. "The Transition from Byzantine to Post-Byzantine Bookbindings: A Statistical Analysis of Some Crucial Changes." In *Book and Paper Conservation*, edited by Jedert Vodopivec Tomažič, 2:12–29. Ljubljana: Archiv Republike Slovenije, 2016.

——. "Twined Endbands in the Bookbinding Traditions of the Eastern Mediterranean." In *Proceedings of the International Conference Men and Books: From Micro-Organism to Mega-Organisms*, St. Hypollit, Austria, 28 April–1 May 2014. Forthcoming.

Boud'hors, Anne. "Copie et circulation des livres dans la region thébaine (VIIe–VIIIe siecle)." In *"Et maintenant ce ne sont plus que des villages . . .": Thèbes et sa région aux époques hellenistique, romain et byzantine*, edited by Alain Delattre and Paul Heilporn, 149–161. Brussels: Association Égyptologique Reine Elizabeth, 2008.

Boud'hors, Anne, and C. Heurtel. "The Coptic Ostraca from the Tomb of Amenemope." *Egyptian Archaeology* 20 (2002): 7–9.

Bowersock, G. W., Peter Brown, and Oleg Grabar, eds. *Late Antiquity: A Guide to the Postclassical World*. Cambridge, Mass.: Belknap Press of Harvard University Press, 2000.

Bowman, Alan Keir, John David Thomas, James Noel Adams, and Richard L. Tapper. *Vindolanda: The Latin Writing Tablets*. Britannia Monograph 4. London: Society for the Promotion of Roman Studies, 1983.

Boyaval, Bernard. "Le cahier scolaire d'Aurèlios Papnouthion." In *Zeitschrift für Papyrologie und Epigraphik* 17 (1975): 225–235.

Brown, Peter Robert Lamont. *The World of Late Antiquity: AD 150–750*. 1971. London: Thames and Hudson, 2006.

Bucking, Scott. *Practice Makes Perfect: P. Cotsen-Princeton 1 and the Training of Scribes in Byzantine Egypt*. Los Angeles: Cotsen Occasional Press, 2011.

Burke, Peter. *Eyewitnessing: The Uses of Images as Historical Evidence*. Ithaca, N.Y.: Cornell University Press, 2001.

Burnham, Dorothy K. "Coptic Knitting: An Ancient Technique." *Textile History* 3, no. 1 (1972): 116–124.

Cameron, Averil. *The Mediterranean World in Late Antiquity AD 395–600*. New York: Routledge, 2003.

Capasso, Mario. "Le tavolette della Villa dei Papiri ad Ercolano." In *Les tablettes à écrire de l'antiquité à l'époque moderne*, edited by Élizabeth Lalou, 221–230. Bibliologia 12. Turnhout, Belgium: Brepols, 1992.

Cardon, Dominique, Hélène Cuvigny, and Dany Nadal. "De pied en cap: A Shoe from Dios and a Hat from Domitianè/Kainè Latomia in the Eastern Desert of Egypt." In *Dress Accessories of the 1st Millennium AD from Egypt: Proceedings of the 6th Conference of the Research Group "Textiles from the Nile Valley," Antwerp, 2–3 October 2009*, edited by Antoine De Moor and Cäcilia Fluck. Tielt, Belgium: Lannoo, 2011.

Cauderlier, Patrice. "Quatre cahiers scolaires (Musée du Louvre): Présentation et problems annexes." In *Les débuts du codex*, edited by Alain Blanchard, 43–59. Bibliologia 9. Turnhout, Belgium: Brepols, 1989.

——. "Les tablettes grecques d'Égypte: Inventaire." In *Les tablettes à écrire de l'antiquité à l'époque moderne*, edited by Élisabeth Lalou, 63–94. Bibliologia 12. Turnhout, Belgium: Brepols, 1992.

Cavallo, Guglielmo. *Libri, edittori e publico nel mondo antico: Cuida storica e critica*. Bari: Biblioteca Universale Laterza, 1975.

——. "Le tavolette come supporto della scrittura: Qualche testimonianza indiretta." In *Les tablets à écrire de l'antiquité à l'époque moderne*, edited by Élisabeth Lalou, 97–104. Bibliologia 12. Turnhout, Belgium: Brepols, 1992.

Clark, Gillian, "City of Books: Augustine and the Worlds as Text." In *The Early Christian Book*, edited by William E. Klingshirn and Linda Safran, 117–138. Washington, D.C.: Catholic University of America Press, 2007.

Clarkson, Christopher. "Some Representations of the Book and Book-Making, from the Earliest Codex Forms through Jost Amman." In *The Bible as Book: The Manuscript Tradition*, edited by John L. Sharpe III and Kimberly Van Kampen, 197–203. New Castle, Del.: Oak Knoll, in association with the Scriptorium, Center for Christian Antiquities, 1998.

Clarysse, Willy, and Katelijn Vandorpe. "Information Technologies: Writing, Book Production and the Role of Literacy." In *The Oxford Handbook of Engineering and Technology in the Classical World*, edited by John Peter Oleson, 716–739. Oxford: Oxford University Press, 2008.

Claßen-Büttner, Ulrike. *Nalbinding: What in the World Is That? History and Technique of an Almost Forgotten Handicraft*. Norderstedt, Germany: Books on Demand, 2015.

Collingwood, Peter. *The Techniques of Tablet Weaving*. London: Faber and Faber, 1982. Reprint, McMinnville, Ore.: Robin and Russ Handweavers, 2002.

Comparato, Frank E. *Books for the Millions: A History of the Men Whose Methods and Machines Packaged the Printed Word*. Harrisburg, Pa.: Stackpole, 1971.

Cribiore, Raffaella. *Writing, Teachers, and Students in Graeco-Roman Egypt*. American Studies in Papyrology 36. Atlanta: Scholars Press, 1996.

Curzon, Robert. *Visits to Monasteries of the Levant*. London: G. Newnes. 1897.

Degni, Paola. *Usi delle tavolette lignee e cerate del mondo greco e romano*. Messina: Sicania, 1998.

De Jonghe, Daniël, Sonja Daemen, Marguerite Rassart-Debergh, Antoine De Moor, and Bruno Overlaet. *Ancient Tapestries of the R. Pfister Collection in the Vatican Library*. Rome: Biblioteca Apostolica Vaticana, 1999.

Del Corso, Lucio, and Rosario Pintaudi. "Papiri letterari dal Museo Egizio del Cairo e una copertina di codice da Antinoupolis." *Papyrologica Florentina* 44 (2015): 3–29.

De Moor, Antoine, Mark Van Strydonck, Mathieu Boudin, Ina Vanden Berghe, Dominique Bénazeth, and Cäcilia Fluck. "Radiocarbon Dating and Colour Patterns of Late Roman to Early Medieval Leather Shoes and Sandals from Egypt." In *Drawing the Threads Together: Textiles and Footwear of the 1st Millennium AD from Egypt*, edited by Antoine De Moor, Cäcilia Fluck, and Petra Linscheid, 164–173. Tielt, Belgium: Lannoo, 2013.

Depuydt, Leo. *Catalogue of Coptic Manuscripts in the Pierpont Morgan Library*. 2 vols. Leuven: Uitgeverij Peeters, 1993.

Dergham, Youssef, and François Vinourd. "Les reliures syriaques: Essai de caractérisation par comparaison avec les reliures byzantines et arméniennes." In *Manuscripta Syriaca: Des sources de première main*, edited by Françoise Briquel-Chatonnet and Muriel Debié, 271–304. Cahiers d'Études Syriaques 4. Paris: Geuthner, 2015.

Déroche, François. "Quelques reliures médiévales de provenance Damascaine." *Revue des Études Islamiques* 54 (1986): 85–99.

Di Bella, Marco. "An Attempt at a Reconstruction of Early Islamic Bookbinding: The Box Binding." In *Care and Conservation of Manuscripts 12: Proceedings of the 12th Seminar on the Conservation of Manuscripts held at the University of Copenhagen 14th–16th October 2009*, edited by M. J. Driscoll, 99–115. Copenhagen: Museum Tusculanum Press, University of Copenhagen, 2011.

——."From Box Binding to Envelope-Flap Binding: The Missing Link in Transitional Islamic Bookbinding." In *Suave Mechanicals III*, edited by Julia Miller, Ann Arbor, Mich.: The Legacy Press, 2016: 264–279.

Doresse, Jean. "Les reliures des manuscrits gnostiques coptes découverts à Khénoboskion." *Revue d' Égyptologie* 13 (1961): 27–49.

Doresse, Jean, and Togo Mina. "Nouveaux textes gnostiques coptes découverts en Haute-Égypte: La bibliothèque de Chenoboskion." *Vigiliae Christianae* 3, no. 3 (1949): 129–141.

Dreibholz, Ursula. "Some Aspects of Early Islamic Bookbindings from the Great Mosque of Sana'a, Yemen." In *Scribes et manuscrits du Moyen-Orient*, edited by François Déroche and Francis Richard, 15–34. Paris: Bibliothèque Nationale de France, 1997.

Edgerton, David. "From Innovation to Use: Ten Eclectic Theses on the Historiography of Technology." *History and Technology* 16, no. 2 (1999): 111–136.

Eliot, Simon, and Jonathan Rose, eds. *A Companion to the History of the Book*. Malden, Mass.: Wiley-Blackwell, 2007.

Emery, Irene. *The Primary Structure of Fabrics: An Illustrated Classification*. Washington, D.C.: Textile Museum, 1966.

Emmel, Stephen. "The Christian Book in Egypt: Innovation and the Coptic Tradition." In *The Bible as Book: The Manuscript Tradition*, edited by John L. Sharpe III and Kimberly Van Kampen, 35–43. New Castle, Del.: Oak Knoll Press, in association with the Scriptorium, Center for Christian Antiquities, 1998.

———. "The 'Coptic Gnostic Library of Nag Hammadi' and the Faw Qibli Excavations." In *Christianity and Monasticism in Upper Egypt*, vol. 2: *Nag Hammadi—Esna*, edited by Gawdat Gabra and Hany N. Takla, 33–43. Cairo: American University in Cairo Press, 2010.

Emmel, Stephen, and Cornelia Eva Römer. "The Library of the White Monastery in Upper Egypt." In *Spätantike Bibliotheken: Leben und Lesen in den frühen Klöstern Ägyptens, Österreichische Nationalbibliothek*, edited by Harald Froschauer and Cornelia Römer, 5–24. Vienna: Phoibos, 2008.

Ettinghausen, Richard. "Near Eastern Book Covers and Their Influence on European Bindings: A Report on the Exhibition 'History of Bookbinding' at the Baltimore Museum of Art, 1957–58." *Ars Orientalis* 3 (1959): 113–131.

Federici, Carlo, and Konstantinos Houlis. *Legature bizantine vaticane*. Rome: Palombi, 1988.

Fluck, Cäcilia, and H. Froschauer. "Dress Accessories from Antinoupolis: Finds from the Northern Necropolis." In *Dress Accessories of the 1st Millennium AD from Egypt: Proceedings of the 6th Conference of the Research Group "Textiles from the Nile Valley," Antwerp, 2–3 October 2009*, edited by Antoine De Moor and Cäcilia Fluck, 54–69. Tielt, Belgium: Lannoo, 2011.

Fluck, Cäcilia, Petra Linscheid, and Susanne Merz. *Textilien aus Ägypten, Teil 1: Textilier aus dem Vorbesitz von Theodor Graf, Carl Schmidt und dem Ägyptischen Museum Berlin*. Wiesbaden: Reichert, 2000.

Forbes, R. J. "Footwear in Classical Antiquity." In *Studies in Ancient Technology*, 2nd ed., 5:58–63. Leiden: Brill, 1966.

Frauberger, Heinrich. *Antike und frühmittelalterliche Fussbekleidungen aus Achmim-Panopolis*. Düsseldorf, 1896.

Frost, Gary. "Adoption of the Codex Book: Parable of a New Reading Mode." *The Book and Paper Group Annual* 17 (1998). Available at http://cool.conservation-us.org/coolaic/sg/bpg/annual/v17/bp17-10.html.

Gacek, Adam. "Arabic Bookmaking and Terminology as Portrayed by Bakr al-Ishbili in His *Kitāb al-taysir fi sinā 'at al-tasfir*." *Manuscripts of the Middle East* 5 (1990–91): 106–113.

Gamble, Harry Y. *Books and Readers in the Early Church: A History of Early Christian Texts*. New Haven, Conn.: Yale University Press, 1995.

Garrucci, P. Raffaele. *Storia della arte cristiana nei primi otto secoli della chiesa*. Vols. 1–6. Prato: Guasti, 1873–1881.

Gaskell, Philip. *A New Introduction to Bibliography*. Oxford: Clarendon, 1972.

Gast, Monika. "A History of Endbands Based on a Study by Karl Jäckel." *The New Bookbinder: Journal of Designer Bookbinders* 3 (1983): 42–58.

Geijer, Agnes. *Birka III: Die Textilfunde aus den Gräbern*. Uppsala, Sweden: Almqvist and Wiksells, 1938.

Godlewski, Włodzimierz. "Naqlun: Excavations, 2002." *Polish Archaelogy in the Mediterranean* 14 (2003): 163–171.

Göpfrich, Jutta, Nina Frankenhauser, and Katharina Mackert. *Wettlauf mit der Vergänglichkeit: A*

Race against Transience. Offenbach, Germany: Deutsches Ledermuseum / Schuhmuseum Offenbach, 2012.

Gottlieb, Theodor. *Bucheinbände: Auswahl von technisch und geschichtlich bemerkenswerten Stücken*. Vienna: Anton Schroll, 1910.

Greene, Kevin. "Historiography and Theoretical Approaches." In *The Oxford Handbook of Engineering and Technology in the Classical World*, edited by John Peter Oleson, 62–90. Oxford: Oxford University Press, 2008.

———. "How Was Technology Transferred in the Western Provinces?" In *Current Research on the Romanization of the Western Provinces*, edited by Mark Wood and Francisco Queiroga, 101–105. Oxford: British Archaeological Reports, 1992.

Greenfield, Jane, and Jenny Hille. *Headbands: How to Work Them*. 3rd ed. New Castle, Del.: Oak Knoll, 2017.

Grohmann, Adolf, and Thomas W. Arnold. *Denkmäler islamischer Buchkunst*. Florence: Pantheon, 1929.

Guidoti, Maria Cristina, ed. *I tessuti del Museo Egizio di Firenze*. Florence: Giunti, 2009.

Gulácsi, Zsuzsanna. *Medieval Manichaean Book Art: A Codicological Study of Iranian and Turkic Illuminated Book Fragments from 8th–11th Century East Central Asia*. Leiden: Brill, 2005.

Haines-Eitzen, Kim. *The Gendered Palimpsest: Women, Writing, and Representation in Early Christianity*. New York: Oxford University Press, 2012.

———. "'Girls Trained in Beautiful Writing': Female Scribes in Roman Antiquity and Early Christianity." *Journal of Early Christian Studies* 6, no. 4 (1998): 629–646.

———. *Guardians of Letters: Literacy, Power, and the Transmitters of Early Christian Literature*. New York: Oxford University Press, 2000.

Hald, Margrethe. *Ancient Danish Textiles from Bogs and Burials: A Comparative Study of Costume and Iron Age Textiles*. Copenhagen: National Museum of Denmark, 1980.

Hansen, Egon H. "Nålebinding: Definition and Description." In *Textiles in Northern Archaeology: NESAT III—Textile Symposium in York, 6–9 May 1987*. North European Symposium for Archaeological Textiles Monograph 3. London: Archetype, 1990.

Haskell, Francis. *History and Its Images: Art and the Interpretation of the Past*. New Haven, Conn.: Yale University Press, 1993.

Hatlie, Peter. *The Monks and Monasteries of Constantinople, ca. 350–850*. Cambridge: Cambridge University Press, 2007.

Hemingway, Séan. "Statuette of an Artisan." Catalogue entry in *Power and Pathos: Bronze Sculpture of the Hellenistic World*, edited by Jens M. Daehner and Kenneth Lapatin, 262–263. Los Angeles: J. Paul Getty Museum, 2015.

Heurtrel, Chantal. "Les prédécesseurs de Frange: L'occupation de TT29 au VIIe siècle." In *Études coptes X: Douzième journée d'études*, edited by A. Boud'hors and C. Louis, 167–178. Paris: De Boccard, 2008.

———. "Que fait Frange dans la cour de la tombe TT29?" *Étude coptes VIII: Dixième journée d'études*, edited by Christian Canuyer, 177–204. Lille: Association Francophone de Coptologie, 2003.

Hobson, Geoffrey Dudley. "Some Early Bindings and Binders' Tools." *Library*, 4th ser., 19 (1938): 202–249.

Hollenback, Kacy L., and Michael Brian Schiffer. "Technology and Material Life." In *The Oxford Handbook of Material Culture Studies*, edited by Dan Hicks and Mary C. Beaudry, 313–332. Oxford: Oxford University Press, 2010.

Hope, Colin A., and K. A. Worp. "Miniature Codices from Kellis." *Mnemosyne* 59, no. 2 (2006): 226–258.

Howard, Margaret. "Technical Description of the Ivory Writing-Boards from Nimrud." *IRAQ* 17, no. 1 (Spring 1955): 14–20.

Humphrey, John William, John Peter Oleson, and Andrew N. Sherwood. *Greek and Roman Technology: A Sourcebook; Annotated Translations of Greek and Latin Texts and Documents*. New York: Routledge, 1998.

Hurtado, Larry W. *The Earliest Christian Artifacts: Manuscripts and Christian Origins*. Grand Rapids, Mich.: Eerdmans, 2006.

Ibscher, Hugo. "Von der Papyrusrolle zum Kodex." *Archiv für Buchbinderei* (1920): 20, 21–23, 28, 33–35, 38–40.

Janssen, Rosalind M. "Soft Toys from Egypt." *Journal of Roman Archaeology* (October 1996); supp. ser. 19, *Archaeological Research in Roman Egypt*, edited by Donald M. Bailey, 231–239.

Jerome, Saint. *Select Letters of St. Jerome.* Translated by F. A. Wright. 1933. Cambridge, Mass.: Harvard University Press, 1991. Available at https://archive.org/details/selectlettersofs 00jerouoft.

Johnston, William M., and Christopher Kleinhenz, eds., *Encyclopedia of Monasticism.* Vols. 1 and 2. Chicago: Fitzroy Dearborn, 2000.

Jones, Heather Rose [Tangwystyl verch Morgant Glasvryn, pseud.]. "A Historical Chain." Is This Stitch Period? series, no. 6. Available at http://www.wkneedle.org/a-historical-chain/.

Kalligerou, Maria. "Tenth-Century Georgian Manuscripts in the Library of the St Catherine's Monastery, Sinai." In *Care and Conservation of Manuscripts 11: Proceedings of the Eleventh International Seminar Held at the University of Copenhagen 24th–25th April 2008*, edited by Matthew James Driscoll and Ragnheiður Mosesdóttir, 151–178. Copenhagen: Museum Tusculanum Press, 2009.

Kebabian, John S. "The Binding of the Glazier Manuscript of the Acts of the Apostles (IVth or IV/Vth century)." In *Homage to a Bookman: Essays on Manuscripts, Books and Printing Written for Hans P. Kraus on His 60th Birthday*, 25–29. Berlin: Mann, 1967.

Kenyon, Frederic G. *The Chester Beatty Biblical Papyri: Descriptions and Texts of Twelve Manuscripts on Papyrus of the Greek Bible.* London: Walker, 1933.

Knötzele, Peter. *Auf Schusters Rappen: Römisches Schuhwerk; Römermuseum Stettfeld.* Karlsruhe: Badisches Landesmuseum, 1996.

Kotsifou, Chrysi. "Bookbinding and Manuscript Illumination in Late Antique and Early Medieval Monastic Circles in Egypt." In *Eastern Christians and Their Written Heritage: Manuscripts, Scribes and Context*, edited by Juan Pedro Monferre-Sala, Herman Teule, and Sofía Torallas Tovar, 213–244. Leuven, Belgium: Peeters, 2012.

——. "Books and Book Production in the Monastic Communities of Byzantine Egypt." In *The Early Christian Book*, edited by William E. Klingshirn and Linda Safran, 48–66. Washington, D.C.: Catholic University of America Press, 2007.

Kouymjian, D. "Armenian Bindings from Manuscript to Printed Book (Sixteenth to Nineteenth Century)," *Gazette du livre médiéval*, no. 49 (2006): 1–14. Available at http://www.palaeographia.org/glm/glm .htm?art=kouymjian.

Lalou, Élisabeth, ed. *Les tablettes à écrire de l'antiquité à l'époque moderne.* Bibliologia 12. Turnhout, Belgium: Brepols, 1992.

Lamacraft, Charles T. "Early Book-Bindings from a Coptic Monastery." *Library*, 4th ser., 20 (1939–1940): 214–233.

Lavan, Luke. "Explaining Technological Change: Innovation Stagnation, Recession and Replacement." In *Technology in Transition A.D. 300–650*, edited by Luke Lavan, Enrico Zanini, and Alexander Sarantis, xv–xl. Leiden: Brill, 2007.

Lavan, Luke, Ellen Swift, and Toon Putzeys, eds. *Objects in Context, Objects in Use: Material Spatiality in Late Antiquity.* Leiden: Brill, 2007.

Leary, T. J. *Martial Book XIV: The Apophoreta, Book 14.* London: Duckworth, 1996.

Lewis, Naphtali, Yigael Yadin, and Jonas C. Greenfield, eds. *The Documents from the Bar Kokhba Period in the Cave of Letters.* Vol. 1. Jerusalem: Israel Exploration Society, 1989.

Lindsay, Jen. "The Edfu Collection of Coptic Books." *New Bookbinder* 21 (2001): 31–51.

Lowden, John. "The Word Made Visible: The Exterior of the Early Christian Book as Visual Argument." In *The Early Christian Book*, edited by William E. Klingshirn and Linda Safran, 13–47. Washington, D.C.: Catholic University of America Press, 2007.

Magoulias, H. J. "Trades and Crafts in the Sixth and Seventh Centuries as Viewed through the Lives of the Saints." *Byzantinoslavica* 37 (1976): 11–35.

Maguire, Eunice Dauterman, Henry Maguire, and Maggie J. Duncan-Flowers. *Art and Holy Powers in the Early Christian House.* Urbana: Krannert Art Museum, University of Illinois at Urbana-Champaign, 1989.

Maguire, Henry. "Garments Pleasing to God: The Significance of Domestic Textile Designs in the Early Byzantine Period." *Dumbarton Oaks Papers* 44 (1990): 215–224.

——. "Magic and Geometry in Early Christian Floor Mosaics and Textiles." *Jahrbuch der Österreichischen Byzantinistik* 44 (1994): 265–274.

——. "Magic and the Christian Image." In *Byzantine Magic*, edited by Henry Maguire, 51–71. Washington, D.C.: Dumbarton Oaks, 1995.

Mannoni, Tiziano. "The Transmission of Craft Techniques according to the Principles of Material Culture: Continuity and Rupture." In *Technology in Transition A.D. 300–650*, edited by Luke Lavan, Enrico Zanini, and Alexander Sarantis, xli–lx. Leiden: Brill, 2007.

Marçais, Georges, and Louis Poinssot. *Objets kairouanais, IXe au XIIIe siècle: Reliures, verreries, cuivres et bronzes, bijoux.* Tunis: Tournier, 1948.

Marichal, Robert. "Les tablettes à écrire dans le monde romain." In *Les tablettes à écrire de l'antiquité à l'époque moderne*, edited by Élisabeth Lalou, 165–185. Bibliologia 12. Turnhout, Belgium: Brepols, 1992.

Martial. *The Epigrams of Martial*, edited by Henry George Bohn. Bohn's Classical Library. London: Bell, 1897.

Mattusch, Carol C. *Pompeii and the Roman Villa: Art and Culture around the Bay of Naples.* Exhibition catalogue. Washington, D.C.: National Gallery of Art; London: Thames and Hudson, 2008.

McCormick, Michael. "The Birth of the Codex and the Apostolic Life-Style." *Scriptorium* 39, no. 1 (1985): 150–158.

McWhirr, Alan. *Roman Crafts and Industries.* Aylesbury, UK: Shire, 1982.

Meeks, Wayne A. *The First Urban Christians: The Social World of the Apostle Paul.* 2nd ed. New Haven, Conn.: Yale University Press, 2003.

Mérat, Amandine. "Étude technique et iconographique d'un ensemble de broderies égyptiennes antiques conservées au musée du Louvre." In *Drawing the Threads Together: Textiles and Footwear of the 1st Millennium AD from Egypt: Proceedings of the 7th Conference of the Research Group "Textiles from the Nile Valley," Antwerp, 7–9 October 2011*, edited by Antoine De Moor, Cäcilia Fluck, and Petra Linscheid, 126–139. Tielt, Belgium: Lannoo, 2013.

Merian, Sylvie L. "The Armenian Bookmaking Tradition in the Christian East: A Comparison with the Syriac and Greek Traditions." In *The Bible as Book: The Manuscript Tradition*, edited by John L. Sharpe III and Kimberly Van Kampen, 205–214. New Castle, Del.: Oak Knoll, in association with the Scriptorium, Center for Christian Antiquities, 1998.

——. "Protection against the Evil Eye? Votive Offerings on Armenian Manuscript Bindings." In *Suave Mechanicals*, edited by Julia Miller, 1:43–93. Ann Arbor, Mich.: Legacy, 2013.

——. "The Structure of Armenian Bookbinding and Its Relation to Near Eastern Bookbinding Traditions." PhD diss., Columbia University, 1993.

Meyer, Elizabeth A. *Legitimacy and Law in the Roman World: Tabulae in Roman Belief and Practice.* Cambridge: Cambridge University Press, 2004.

——. "Roman Tabulae, Egyptian Christians, and the Adoption of the Codex." *Chiron* 37 (2007): 295–347.

——. "Writing Paraphernalia, Tablets, and Muses in Campanian Wall Paintings." *American Journal of Archaeology* 113, no. 4 (2009): 569–597.

Meyer, Wendy. "The Changing Shape of Liturgy: From Earliest Christianity to the End of Late Antiquity." In *Liturgy's Imagined Past/s: Methodologies and Material in the Writing of Liturgical History Today*, edited by Teresa Berger and Bryan Spinks, 271–298. Collegeville, Minn.: Liturgical Press, 2016.

Millard, Alan R. *Reading and Writing in the Time of Jesus.* Sheffield, UK: Sheffield Academic Press, 2001.

Montembault, Véronique. *Catalogue des chaussures de l'antiquité égyptienne.* Paris: Réunion des Musées Nationaux, 2000.

Morey, Charles R. "The Painted Covers of the Washington Manuscript of the Gospels." In *East Christian Paintings in the Freer Collection*, edited by Charles Rufus Morey and Charles Lang Freer, 63–81. New York: Macmillan, 1914.

Musurillo, Herbert. "The Martyrdom of Agape, Irene and Chione." In *The Acts of the Christian Martyrs (Early Christian Texts): Introduction, Texts and Translations.* Oxford: Clarendon Press, 1972.

Nauerth, Claudia. *Die koptischen Textilien der Sammlung Wilhelm Rautenstrauch im Städtischen Museum Simeonstift Trier*. Trier, Germany: Selbstverlag des Städtischen Museums Simeonstift, 1989.

——. "Sandalen, Schuhe und Pantoffeln mit Vergoldung." In *Drawing the Threads Together: Textiles and Footwear of the 1st Millennium AD from Egypt: Proceedings of the 7th Conference of the Research Group "Textiles from the Nile Valley," Antwerp, 7–9 October 2011*, edited by Antoine De Moor, Cäcilia Fluck, and Petra Linscheid, 264–267. Tielt, Belgium: Lannoo, 2013.

Needham, Paul. *Twelve Centuries of Bookbindings 400–1600*. New York: Pierpont Morgan Library / Oxford University Press, 1979.

Nicholls, Matthew. "Parchment Codices in a New Text of Galen." *Greece and Rome* 57, no. 2 (October 2010): 378–386.

Noever, Peter, ed. *Fragile Remnants: Egyptian Textiles of Late Antiquity and Early Islam*. Mak Studies 5. Exhibition catalogue. Vienna: Hatje Cantz, 2005.

Olson, Carl. "The Sacred Book." In *The Oxford Companion to the Book*, edited by Michael F. Suarez and H. R. Woudhuysen, 1:11–23. New York: Oxford University Press, 2010.

O'Neale, Lila M. "Peruvian 'Needleknitting.'" *American Anthropologist*, n.s., 36, no. 3 (July–September 1934): 405–430.

I Papiri Bodmer: Biblioteche, comunità di asceti e cultura letteraria in greco, copto e latino nell'Egitto tardoantico. Monograph section of *Adamantius* 21 (2015): 8–171.

Parani, Maria G. *Reconstructing the Reality of Images: Byzantine Material Culture and Religious Iconography (11th–15th Centuries)*. Boston: Brill, 2003.

Pattie, Thomas S. "The Creation of the Great Codices." In *The Bible as Book: The Manuscript Tradition*, edited by John L. Sharpe III and Kimberly Van Kampel, 61–72. New Castle, Del.: Oak Knoll, in association with the Scriptorium, Center for Christian Antiquities, 1998.

Payton, Robert. "The Ulu Burun Writing Board Set." *Anatolian Studies* 41 (1991): 99–106.

Petersen, Theodore C. "Coptic Bookbindings in the Pierpont Morgan Library," 1948. Typescript, Morgan Library and Museum.

——. "Early Islamic Bookbindings and Their Coptic Relations." *Ars Orientalis* 1 (1954): 41–64.

Petherbridge, Guy. "Sewing Structures and Materials: A Study in the Examination and Documentation of Byzantine and Post-Byzantine Bookbindings." In *Paleografia e codicologia greca: Atti del II colloquio internazionale* (Berlino-Wolfenbüttel, 17–21 ottobre 1983), edited by Dieter Harlfinger and Giancarlo Prato, 1:363–408, 2:201–209. Alessandria, Italy: Edizioni dell'Orso, 1981.

Petrie, W. M. Flinders. *Objects of Daily Use: Illustrated by the Egyptian Collection in University College*. London: British School of Archaeology in Egypt, 1927. Reprint, Encino, Calif.: Malter, 1974.

Pfister, Rodolphe, and Louisa Bellinger. *The Excavations at Dura-Europos* IV, 2: *The Textiles*. New Haven, Conn.: Yale University Press, 1945.

Piccirillo, Michele. *The Mosaics of Jordan*, edited by Patricia M. Bikai and Thomas A. Dailey. Amman: American Center of Oriental Research, 1993.

Pickwoad, Nicholas. *Assessment Manual: A Guide to the Survey Forms to Be Used in St Catherine's Monastery*. London, April 2004. http://www.ligatus.org.uk/sites/default/files/manual20050110.pdf.

——. "Binding." In *The St Cuthbert Gospel: Studies on the Insular Manuscript of the Gospel of John*, edited by Claire Breay and Bernard Meehan, 41–63. London: British Library, 2015.

Pritchard, Frances. *Clothing Culture: Dress in Egypt in the First Millennium AD: Clothing from Egypt in the collection of the Whitworth Gallery*. Manchester, UK: Whitworth Art Gallery, 2006.

——. "A Survey of Textiles in the UK from the 1913–14 Egypt Exploration Fund Season at Antinoupolis." In *Drawing the Threads Together: Textiles and Footwear of the 1st Millennium AD from Egypt: Proceedings of the 7th Conference of the Research Group "Textiles from the Nile Valley," Antwerp, 7–9 October 2011*, edited by Antoine De Moor, Cäcilia Fluck, and Petra Linscheid, 35–55. Tielt, Belgium: Lannoo, 2013.

Puglia, Enzo. *La cura del libro nel mondo antico: Guasti e restauri del rotolo di papiro*. Naples: Liguori, 1997.

Pugliese Carratelli, Giovanni. "L'instrumentum scriptorium nei monumenti pompeiani ed ercolanesi." In *Pompeiana: Raccolta di studi per il secondo centenario degli scavi di Pompei*, 266–278. Naples: Macchiaroli, 1950.

Regemorter, Berthe van. "Le codex relié à l'époque néo-hittite." *Scriptorium* 12, no. 2 (1958): 177–181.

———. "Le codex relié depuis son origine jusqu'au haut moyen-âge." *Le Moyen Âge* (1955): 1–26.

———. "Some Early Bindings from Egypt at the Chester Beatty Library." Chester Beatty Monograph 7. Dublin: Hodges Figgis, 1958.

Reuterswärd, Patrik. "The Christian Use of the Tetragram." In *World Art: Themes of Unity in Diversity; Acts of the XXVth International Congress of the History of Art*, edited by Irvin Lavin, 1:219–222. University Park: Pennsylvania State University Press, 1989.

———. "The Forgotten Symbols of God." In *The Visible and Invisible in Art: Essays in the History of Art*, 57–136. Vienna: IRSA, 1991.

Roberts, Colin H., and T. C. Skeat. *The Birth of the Codex*. New York: Oxford University Press, 1983.

Rousseau, Philip. *Pachomius: The Making of a Community in Fourth-Century Egypt*. Rev. ed. Berkeley: University of California Press, 1999.

Ruprechtsberger, Erwin Maria. *Syrien: Von den Aposteln zu den Kalifen*. Mainz, Germany: Zabern, 1993.

Rutt, Richard. *A History of Hand Knitting*. Loveland, Colo.: Interweave, 1987.

Sarris, Nikolas. "Classification of Finishing Tools in Greek Bookbinding: Establishing Links from the Library of St Catherine's Monastery, Sinai, Egypt." PhD thesis, Camberwell College of Arts, University of the Arts, London, 2010.

Schachner, Lukas Amadeus. "Book Production in Early Byzantine Monasteries." In *St Catherine's Monastery at Mount Sinai: Its Manuscripts and Their Conservation; Papers Given in Memory of Professor Ihor Ševčenko*, edited by Cyril A. Mango, 21–30. London: Saint Catherine Foundation, 2011.

Scheller, Robert Walter Hans Peter. *Exemplum: Model-Book Drawings and the Practice of Artistic Transmission in the Middle Ages (ca. 900–ca. 1470)*. Translated by Michael Hoyle. Amsterdam: Amsterdam University Press, 1995.

Scherer, Jean. "Liste d'objets à acheter à Alexandrie." In *Les Papyrus Fouad Ier, Publications de la Société Fouad I de Papyrologie, Textes et Documents, nos. 1–89*, edited by A. Bataille, O. Guéraud, P. Jouguet, N. Lewis, H. Marrou, J. Schérer, and W. G. Waddell, 155–156. Cairo: Imprimerie de l'Institut Français d'Archéologie Orientale, 1939.

Schmidt-Colinet, Andreas. "Les deux carrés entrelacés inscrits dans un cercle: De la signification d'un ornement géométrique." In Annemarie Stauffer, *Textiles d'Égypte de la collection Bouvier: Antiquité tardive, période copte, premiers temps de l'Islam*, 21–34. Exhibition catalogue. Fribourg: Musée d'Art et d'Histoire, 1991.

Schmidt-Colinet, Andreas, Annemarie Stauffer, and Khaled Al-As'ad, *Die Textilien aus Palmyra: Neue und alte Funde*. Mainz am Rhein, Germany: Zabern, 2000.

Schoeser, Mary. *World Textiles: A Concise History*. London: Thames and Hudson, 2003.

Schubart, Wilhelm. *Das Buch bei den Griechen und Römern: Eine Studie aus der Berliner Papyrussammlung*. Berlin: Reimer, 1907.

Seiler-Baldinger, Annemarie. *Textiles: A Classification of Techniques*. Washington, D.C.: Smithsonian Institution Press, 1994.

Sharpe, John L. "Dakhleh Tablets and Some Codicological Considerations." In *Les tablettes à écrire de l'antiquité à l'époque moderne*, edited by Élisabeth Lalou, 127–148. Bibliologia 12. Turnhout, Belgium: Brepols, 1992.

———. "The Earliest Bindings with Wooden Board Covers: The Coptic Contribution to Binding Structures." In *International Conference on Conservation and Restoration of Archival and Library Materials* (Erice, April 22–29, 1996), edited by Carlo Federici, Paola F. Munafò, and Daniela Costantini, 455–478. Palermo: Palumbo, 1999.

———. "Wooden Books and the History of the Codex: Isocrates and the Farm Account, Evidence from the Egyptian Desert." In *Roger Powell the Compleat Binder, Liber Amicorum*, edited by John L. Sharpe, 107–128. Bibliologia 14. Turnhout, Belgium: Brepols, 1996.

Shore, Arthur Frank. "Fragment of a Decorated Leather Binding from Egypt." *British Museum Quarterly* 36, no. 1/2 (Autumn 1971): 19–23.

Sirat, Colette. "Le codex de bois." In *Les débuts du codex: Actes de la journée d'étude organisée à Paris les 3 et 4 juillet 1985 par l'Institut de Papyrologie de la Sorbonne et l'Institut de Recherche et d'Histoire des Textes*, edited by Alain Blanchard, 37–40. Bibliologia 9. Turnhout, Belgium: Brepols, 1989.

Skeat, Theodore C. "Early Christian Book-Production: Papyri and Manuscripts." In *The Cambridge History of the Bible*, vol. 2: *The West from the Fathers to the Reformation*, edited by G. W. H. Lampe, 54–79, 512–513. Cambridge: Cambridge University Press, 1969.

———. "Especially the Parchments: A Note on 2 Timothy 4.13," *Journal of Theological Studies* 30 (1979): 173–177.

———. "The Length of the Standard Papyrus Roll and the Cost-Advantage of the Codex." *Zeitschrift für Papyrologie und Epigraphik* 45 (1982): 169–175.

———. "The Origin of the Christian Codex." *Zeitschrift für Papyrologie und Epigraphik* 102 (1994): 263–268.

Speidel, M. Alexander. *Die römischen Schreibtafeln von Vindonissa: Lateinische Texte des militärischen Alltags und ihre geschichtliche Bedeutung*. Brugg, Switzerland: Pro Vindonissa, 1996.

Spitzmueller, Pamela. "A Trial Terminology for Describing Sewing through the Fold." *Paper Conservator* 7 (1982): 44–46.

Stanton, Graham N. *Jesus and Gospel*. Cambridge: Cambridge University Press, 2004.

Stauffer, Annemarie. "Cartoons for Weavers from Graeco-Roman Egypt." In *Archaeological Research in Roman Egypt: Proceedings of the Seventeenth Classical Colloquium of the Department of Greek and Roman Antiquities, British Museum, Held on 1–4 December, 1993*, edited by Donald M. Bailey, 223–230. Ann Arbor, Mich.: Journal of Roman Archaeology, 1996.

Stevick, Robert D. "St. Cuthbert Gospel Binding and Insular Design." *Artibus et Historiae* 8, no. 15 (1987): 9–19.

Strong, Donald Emrys, and David Brown. *Roman Crafts*. New York: New York University Press, 1976.

Strzygowski, Josef. *Koptische Kunst: Catalogue général des antiquités égyptiennes du Musée du Caire*. Vienna: Adolf Holzhausen, 1904.

Available at https://archive.org/details/koptischekunst00strz.

Symington, Dorit. "Late Bronze Age Writing-Boards and Their Uses: Textual Evidence from Anatolia and Syria." *Anatolian Studies* 41 (1991): 111–123.

Szirmai, J. A. *The Archaeology of Medieval Bookbinding*. Aldershot, UK: Ashgate, 1999.

———. "Wooden Writing Tablets and the Birth of the Codex." *Gazette du livre médiévale*, no. 17 (Autumn 1990): 31–32.

Talbot, A-M. "Monasticism." In *The Oxford Dictionary of Byzantium*, edited by Alexander P. Kazhdan, 2:1392–1394. New York: Oxford University Press, 1991.

Tefnin, R. "A Coptic Workshop in a Pharaonic Tomb." *Egyptian Archaeology* 20 (2002): 6.

Tímár-Balázsy, Ágnes, and Dinah Eastop. *Chemical Principles of Textile Conservation*. 1998. Oxford: Butterworth-Heinemann, 2005.

Timbie, Janet. "Egypt." In *Encyclopedia of Monasticism*, edited by William M. Johnston, 1:432–435. Chicago: Fitzroy Dearborn, 2000.

Les tranchefiles brodées: Étude historique et technique. Paris: Bibliothèque Nationale de France, 1989.

Trilling, James. "Medieval Interlace Ornament: The Making of a Cross-Cultural Idiom." *Arte Medievale*, 2nd ser., 9, no. 2 (1995): 59–86.

———. *The Roman Heritage: Textiles from Egypt and the Eastern Mediterranean 300 to 600 AD*. Washington, D.C.: Textile Museum, 1982.

Turnau, Irena. *History of Knitting before Mass Production*. Translated by Agnieska Szonert. Warsaw: Institute of the History of Material Culture, Polish Academy of Sciences, 1991.

Turner, Eric G. *The Typology of the Early Codex*. Philadelphia: University of Pennsylvania Press, 1977.

Türr, Katja Marina. *Eine Musengruppe hadrianischer Zeit: Die sogenannten Thespiaden*. Monumenta Artis Romana 10. Berlin: Mann, 1971.

Ulrich, Roger B. *Roman Woodworking*. New Haven, Conn.: Yale University Press, 2007.

Van Driel-Murray, C. "Leatherwork and Skin Products." In *Ancient Egyptian Materials and Technology*, edited by Paul T. Nicholson and Ian Shaw, 299–319. Cambridge: Cambridge University Press, 2000.

——. "Vindolanda and the Dating of Roman Footwear." *Britannia* 32 (2001): 185–197.

Van Elderen, Bastiaan. "Early Christian Libraries." In *The Bible as Book: The Manuscript Tradition*, edited by John L. Sharpe III and Kimberly Van Kampen, 45–59. New Castle, Del.: Oak Knoll, in association with the Scriptorium, Center for Christian Antiquities, 1998.

Van Haelst, Joseph. "Les origines du codex." In *Les débuts du codex*, edited by Alain Blanchard, 13–35. Bibliologia 9. Turnhout, Belgium: Brepols, 1989.

Van Strydonck, Mark, Antoine De Moor, and Dominique Bénazeth. "14C Dating Compared to Art Historical Dating of Roman and Coptic Textiles from Egypt." *Radiocarbon* 46, no. 1 (2004): 231–244.

Veilleux, Armand, trans. *Pachomian koinonia*. Vol. 1: *The Life of Saint Pachomius and His Disciples*; vol. 2: *Pachomian Chronicles and Rules*. Kalamazoo, Mich.: Cistercian Publications, 1980–1981.

Veldmeijer, André J. *Sailors, Musicians and Monks: The Leatherwork from Dra' Abu el Naga (Luxor, Egypt)*. Leiden: Sidestone, 2017.

——. *Sandals, Shoes and Other Leatherwork from the Coptic Monastery Deir el-Bachit: Analysis and Catalogue*. Leiden: Sidestone, 2011.

——. "Studies of Ancient Egyptian Footwear: Technological Aspects. Pt. 16: Leather Open Shoes." *British Museum Studies in Ancient Egypt and Sudan* 11 (2009): 1–10.

——. *Tutankhamun's Footwear: Studies of Ancient Egyptian Footwear*. Leiden: Sidestone, 2011.

Veldmeijer, André J., and Salima Ikram. *Catalogue of the Footwear in the Coptic Museum (Cairo)*. Leiden: Sidestone, 2014.

Vogelsang-Eastwood, Gillian, "Textiles." In *Ancient Egyptian Materials and Technology*, edited by Paul T. Nicholson and Ian Shaw, 268–298. Cambridge: Cambridge University Press, 2000.

Von Knötzele, Peter. *Römische Schuhe: Luxus an den Füßen*. Schriften des Limesmuseums Aalen, no. 59. Stuttgart: Theiss, 2007.

Waelkens, Marc. *Die kleinasiatischen Türsteine: Typologische und epigraphische Untersuchungen der kleinasiatischen Grabreliefs mit Scheintür*. Mainz am Rhein: Zabern, 1986.

Warner Dendel, Esther. *African Fabric Crafts: Sources of African Design and Technique*. New York: Taplinger, 1974.

Waterer, J. W. "Leatherwork." In *Roman Crafts*, edited by Donald E. Strong and David Brown, 179–194. New York: New York University Press, 1976.

Wendrich, Willemina Z. "Basketry." In *Ancient Egyptian Materials and Technology*, edited by Paul T. Nicholson and Ian Shaw, 254–267. Cambridge: Cambridge University Press, 2000.

——. *The World according to Basketry: An Ethno-Archaeological Interpretation of Basketry Production in Egypt*. Leiden: Research School of Asian, African and Amerindian Studies (CNWS), Universiteit Leiden, 1999.

Wild, John Peter. *Textiles in Archaeology*. Princes Risborough, Aylesbury, Bucks: Shire, 1988.

Winlock, Herbert Eustis, and W. E. Crum. *The Monastery of Epiphanius at Thebes*. Vol. 2. Edited by Hugh G. Evelyn-White. New York: Metropolitan Museum of Art, 1926.

Wipszycka, Ewa. "Les aspects économiques de la vie de la communauté des Kellia." In *Le site monastique des Kellia: Sources historiques et exploration archéologiques*, edited by Philippe Bridel, 117–144. Geneva: Mission Suisse d'Archéologie Copte de l'Université de Genève, 1986. Reprinted in Ewa Wipsyzycka, *Étude sur le christianisme dans l'Égypte de l'antiquité tardive*, 337–362. Rome: Institutum Patristicum Augustinianum, 1996.

——. "Resources and Economic Activities of the Egyptian Monastic Communities (4th–8th Century)." *Journal of Juristic Papyrology* 41 (2011): 159–263.

Wulff, Oskar Konstantin. *Altchristliche und mittelalterliche byzantinische und italienische Bildwerke*. Vol. 1. Berlin: Reimer, 1909.

Yadin, Yigael, Hannah Cotton, and Andrew Gross. *The Documents from the Bar Kokhba Period in the Cave of Letters*. Vol. 2: *Hebrew, Aramaic, and Nabatean-Aramaic Papyri*. Jerusalem: Israel Exploration Society, 2002.

Yamauchi, Edwin M. "The Nag Hammadi Library." *Journal of Library History* 22, no. 4 (1987): 425–441.

Index

papyrus, xiii, 1–2, 35–37, 39–40, 46, 80 laminate, 40, 69, 80, 106
parchment, xiii, 1–2, 8, 32n9, 35–37, 80, 109, 112
Parry, Milman, xii
Paulus, Iulius, 2
pens, 2, 32n9
Pesynthios, 155
Petersen, Theodore C., 14, 47, 53, 60–61, 72, 78n10, 89, 100, 149
Plato, xii–xiii
Pompeii, 1, 2, 26, 30, 33n18, 34n24, 67n22
post-Byzantine bookbindings, 50, 85, 141
protective and apotropaic devices, 128–129
pugillares, 23, 35

Q

Quintilian, 32n9
quires, 9, 17n22

R

Ragyndrudis codex, 66n8, 112
Ravenna, mosaics from, 65, 83, 134, 135
Regemorter, Berthe van, 12
Reuterswärd, Patrick, 121, 126–127, 132n46
Ricard, Prosper, 86
rolls, xiii, 1–3, 12, 153

S

sandals, 101, 117, 138, 141, 146n2,6
Sappho, 30
School tablets, 24–27
Scott, James C., xii
scrolls. *See* rolls
sewing, 8, 9, 13, 14, 28–30, 49–63, 153–154; chain stitch, 53–54; cross-knit looping, 54–59, 71, 75; double-sequence, 66n6; herringbone, 50, 60; loop-stitch, 52, 61, 68n37; stab, 40–41, 42n16, 61–63; stations, 51–52, 61–62, 66nn7–8, 84; unsupported vs. supported, 49–51
Sheide codex (Princeton University Library, Sheide 144), 84

Shepherd of Hermas, 7
shoemaking, xiv, 6, 13–14, 98, 101–103, 108, 114–117, 142, 154
sillybus, 2
stylus, 3, 23, 31, 100
Sisters of Thessaloniki, martyrdom of, 8
socks, 55–59
Sinai codex. *See* St. Catherine's Monastery codices
spatula, 2
spine lining, 45, 76–77, 79–81
spine strips, 37, 38–40
St. Catherine's Monastery codices, 15, 17n12, 47, 48n8, 81, 82n3,4, 84, 86, 89, 90, 92, 94–95, 100, 101, 106–107, 106, 120, 127, 141, 149, 150
St. Cuthbert Gospel, 47, 50–51, 66n8, 78n9, 87
Szirmai, János A., 13, 26, 46–47, 85

T

tabulae. *See* wooden tablets
tackets, tacketing, 38, 39, 79, 84
tablets, different types of, 24, 25, methods for connecting, 26–30
Theodoros, tablet codex of, 27, 29, 33n20
Tollund Man, 141
textile edging, 86, 96n11
tooling (impressing, stamping), 100–103, 106, 129n5
Trebius Iustus, 2, 3, 31
Trilling, James, 118, 120, 129, 132n44
Turner, Eric G., 13, 18n25, 37
Twining, standard 90–92, split, 93

U

Uluburun diptych, 22

W

wall paintings, 1–2, 17n3, 22, 26, 30–31, 136
wax tablets, 2, 23, 24–25, 26–28, 31, 32n3, 32n9, 32n12, 37
Wipszycka, Ewa, 155
wooden tablets, 1–2, 6, 8, 9, 21–31, 32n4, 36–37, 46, 106, 153

Photography Credits

AA World Travel Library / Alamy Stock Photo: fig. 41 left

Jon Arnold Images Ltd. / Alamy Stock Photo: fig. 102

Georgios Boudalis, figs. 27, 52, 61, 66, 69, 72, 85, 101, 121, 122

bpk Bildagentur / Vorderasiatisches Museum, Staatliche Museen / Art Resource, NY: fig. 114

Catacombe di Napoli: fig. 23

De Agostini Picture Library / Getty Images: fig. 42

Fotografica Foglia © Scala / Art Resource, NY: figs. 1, 17

Getty Images: fig. 41 right

Barry Iverson / Alamy Stock Photo: fig. 125

Lanmas / Alamy Stock Photo: fig. 18

Image copyright © The Metropolitan Museum of Art. Image source: Art Resource, NY: figs. 19, 58, 59, 81, 87, 90, 93, 94, 107, 108, 109

Tony Querrec © RMN-Grand Palais / Art Resource, NY: fig. 13

Rea, Rossella. L'ipogeo Di Trebio Giusto Sulla Via Latina: Scavi E Restauri. Città del Vaticano: Pontificia commissione di archeologia sacra, 2004: fig. 2

Scala / Art Resource: figs. 71, 103

Bruce White: figs. 34, 45, 46; cats. 2, 3, 5, 20, 45

Stamatis Zoumpourtikoudis: cats. 2, 3, 4, 5, 7, 8, 9, 10, 14, 15, 16, 24